Applied Operational Research
with SAS

Applied Operational Research with SAS

Ali Emrouznejad

William Ho

CRC Press
Taylor & Francis Group
Boca Raton London New York

CRC Press is an imprint of the
Taylor & Francis Group, an **informa** business

A CHAPMAN & HALL BOOK

CRC Press
Taylor & Francis Group
6000 Broken Sound Parkway NW, Suite 300
Boca Raton, FL 33487-2742

First issued in paperback 2020

© 2011 by Taylor & Francis Group, LLC
CRC Press is an imprint of Taylor & Francis Group, an Informa business

No claim to original U.S. Government works

ISBN-13: 978-0-367-57681-3 (pbk)
ISBN-13: 978-1-4398-4130-3 (hbk)

Library of Congress Cataloging-in-Publication Data

Emrouznejad, Ali.
  Applied operational research with SAS / Ali Emrouznejad, William Ho.
    p. cm.
  Includes bibliographical references and index.
  ISBN 978-1-4398-4130-3 (hardback)
  1. Operations research--Data processing. 2. SAS (Computer file) I. Ho, William. II. Title.

T57.5.E47 2011
658.4'032--dc23   2011043666

Visit the Taylor & Francis Web site at
http://www.taylorandfrancis.com

and the CRC Press Web site at
http://www.crcpress.com

Contents

Preface

This book was written primarily for SAS (Statistical Analysis System) users in operational research, or simply OR. We use a wide range of OR optimization problems to demonstrate how the SAS/OR® procedures work. The problems include single criterion optimization problems (e.g., transportation, assignment, transshipment, minimum-cost capacitated flow, maximum flow, shortest path, assembly line balancing, and traveling salesman), project management decision problems (e.g., critical path analysis, and program evaluation and review technique), printed circuit board (PCB) assembly problems (e.g., PCB assembly line assignment, PCB component allocation, and PCB component sequencing for pick-and-place machines), and multiple criteria decision-making problems (e.g., analytic hierarchy process, multiple criteria logistics distribution, ordered weighted averaging, data envelopment analysis, and Malmquist productivity index). The emphasis in this book is to use PROC OPTMODEL, which is the most recent development in SAS for optimization problems. The principles of several other procedures are also presented. We believe that this is the first book to extensively and successfully cover the application of SAS/OR® procedures to OR problems.

In addition, this book was written for undergraduate and postgraduate students in OR and those who are interested in mathematical modeling techniques. We formulated the OR problems mentioned earlier as various types of mathematical models, including linear programming, integer linear programming, and goal-programming models. We also show the transformation of the minimax-type formulation into a minimization-type formulation.

This book is organized as follows. Chapter 1 first presents the algorithms and methods for linear programming, integer linear programming, and goal-programming models. The chapter then describes SAS/OR®, the main part of which is optimization, and it consists of several procedures. Then the principles of these procedures are presented in this chapter. Chapters 2 to 9 demonstrate how SAS/OR® procedures can be applied to solve various types of OR problems. Each chapter describes the concept of an OR problem, presents an example of the problem, and explains the specific procedure and its macros for solving the problem to optimality. These macros include the data-handling macro, the model-building macro, and the report-writing macro.

As the authors of this book, we wish to express our sincere gratitude to the SAS Institute Inc., and to the reviewers for their support and helpful comments. Dr. Ali Emrouznejad dedicates this book to his wife Forouzan for her true love and continuous support, and to their children, Pardis, Pedram, and Parham, for their patience. Dr. William Ho is genuinely thankful to his wife

Peggy and his son Andreas for their endless love and concern, and to Dr. Ali Emrouznejad for his invitation to jointly write this book.

We are also indebted to our editor, David Grubbs, project manager, Srikanth Gopaalan, production editor, Rachael Panthier, and others at Taylor & Francis Group, LLC (CRC Press) for their editorial counsel and support during the preparation of this book.

1

Operational Research, Algorithms, and Methods

The term *operations research* (or *operational research* as appears in this book) was introduced in England during World War II when British military leaders ordered scientists to make decisions concerning the optimal use and allocation of limited war material and resources such as radar and bombing. After the war, the success of operational research was extensively recognized.

Operational research is a scientific decision-making tool that involves the use of a mathematical programming model. A mathematical programming model is a mathematical representation of the actual situation that may be used to make better decisions or simply to understand the actual situation better (Winston and Venkataramanan 2003). The common feature that mathematical programming models have is that they all involve optimization (Williams 1999), which includes the minimization of something (e.g., delivery time and production cost) or the maximization of something (e.g., customer service level and profit) under certain constraints (e.g., budget and human resources).

A set of fixed computational rules for solving a particular class of problems or models is known as an *algorithm*. It applies the rules repetitively to the problem or the model, and each iteration moves the solution closer to the optimum. In operational research, there is no algorithm that solves all types of mathematical models. For example, the simplex method is the general method for solving linear programming models, whereas the branch-and-bound algorithm is the general technique for solving integer linear programming models.

In the following sections, attention is confined to the algorithms and methods for the linear programming model, integer linear programming model, and goal programming model. They are discussed because the practical examples, to be examined in Chapters 2 to 9, can be formulated with these types of models.

1.1 Linear Programming

A *linear programming (LP) model* comprises three basic elements: decision variables, objectives, and constraints. A model is defined as LP when the objective function and the constraints involve linear expressions and the decision variables are continuous. The transportation model, to be presented in Section 2.1, is a special class of LP. Comparatively, LP models are given extensive attention in comparison with nonlinear programming models because they are much easier to solve and they have been applied successfully in many contexts, including agriculture, business, economics, environmental studies, government, higher education, logistics, manufacturing, and military planning.

The first step in formulating the LP model is to define the *decision variables*. They can be expressed in any form except nonlinear functions, such as x_1^2 and x_1x_2. Decision variables are the objects that the user needs to determine. For example, the decision variables in the transportation model are the quantities of commodities sent from a set of origins to a set of destinations.

After defining the decision variables, the user has to define an *objective*, which is the goal that they aim to optimize. Some prevalently used objectives include maximization of profit, maximization of workload balance, maximization of efficiency, maximization of customer satisfaction, minimization of cost, minimization of travelling distance, minimization of cycle time, and minimization of vehicles used.

The last elements of LP models are the *constraints*, which are the conditions that the user needs to satisfy. Some of the most common types of constraints used in LP models include customer demands, available workforce, available raw material, production time, available machinery, budget constraints, and subtour elimination constraints.

The general LP model can be formulated as shown in Model 1.1.1.

Model 1.1.1 Standard maximization-type linear programming model

$$\text{Maximize } z = \sum_{j=1}^{n} c_j x_j \tag{1.1.1}$$

subject to

$$\sum_{j=1}^{n} a_{ij} x_j \leq b_i \quad \text{for all } i \tag{1.1.2}$$

$$\text{All } x_j \geq 0$$

In this LP model, z is the optimal solution value, x_j are the decision variables, a_{ij} are the constraint coefficients, b_i are the right-side values, and c_j are the objective function coefficients. Objective function 1.1.1 maximizes the total summation of c_j, while subject to constraint set 1.1.2. An implicit condition on the model is that all decision variables must be positive. Therefore, the nonnegativity constraints $x_j \geq 0$ are added. Generally, the $\leq$ sign is used in the constraint of a maximization-type LP model, whereas the $\geq$ sign is used when an LP model belongs to the minimization type.

Model 1.1.2 Standard minimization-type linear programming model

$$\text{Minimize } z = \sum_{j=1}^{n} c_j x_j \qquad (1.1.3)$$

subject to

$$\sum_{j=1}^{n} a_{ij} x_j \geq b_i \quad \text{for some of } i \qquad (1.1.4)$$

$$\sum_{j=1}^{n} a_{ij} x_j = b_i \quad \text{for some of } i \qquad (1.1.5)$$

$$\text{All } x_j \geq 0$$

Models 1.1.1 and 1.1.2 are standard LP models in which the objective function is either maximization or minimization. There are some occasions in which an LP model may consist of a minimax-type objective function, such as minimization of the maximum production time among several machines. This kind of objective function can be applied to reduce the cycle time or increase the efficiency of a production line. Further information about the minimax-type formulation can be found in Section 7.2.

If the objective function of an LP model is a minimax type or written as shown in Model 1.1.3.

Model 1.1.3 Standard minimax-type linear programming model

$$\text{Minimize } z = \left(\text{Maximum} \sum_{j=1}^{n} c_{ij} x_j \right) \quad \text{for all } i \qquad (1.1.6)$$

subject to

$$\sum_{j=1}^{n} a_{ij} x_j = b_i \quad \text{for all } i \qquad (1.1.7)$$

$$\text{All } x_j \geq 0$$

This can be converted into a conventional LP form by introducing a variable T to represent the objective (Williams 1999). The transformed model can be written as shown in Model 1.1.4.

Model 1.1.4 Transformed minimax-type linear programming model

$$\text{Minimize } z = T \qquad (1.1.8)$$

subject to

$$\sum_{j=1}^{n} c_{ij}x_j - T \leq 0 \quad \text{for all } i \tag{1.1.9}$$

$$\sum_{j=1}^{n} a_{ij}x_j = b_i \quad \text{for all } i \tag{1.1.10}$$

$$\text{All } x_j \geq 0$$

After formulating LP models, we can solve them to optimality. When we attempt to solve the models, one or more of four special cases may arise: infeasibility, unboundedness, redundancy, and alternative optimal solutions. The first special case, *infeasibility*, signifies that there is no solution to the LP model to satisfy all the constraints and the nonnegativity conditions. For example, infeasibility will occur when the amount of resources available is less than the targeted level of production. In this case, the user should revise the model by providing additional resources. The second, *unboundedness*, represents the idea that the solution to the LP model is unbounded. For example, the profits can be maximized infinitely without violating any of the constraints in the maximization-type LP models. Unboundedness implies that the LP model is not formulated properly. In this situation, the user should review the model by ensuring that all necessary constraints have been included. The third special case, *redundancy*, represents the idea that the optimal solution to the LP model will not be changed, even if a particular constraint is left out. This constraint can be regarded as redundant. Redundancy is often not obvious until the model is solved. Once the redundant constraints are found, the user should omit them so that the computational complexity of the model can be reduced. The fourth special case, *alternative optimal solutions*, signifies that more than one solution can yield the optimal value for the objective function. As the case with redundancy, determining whether an LP model has alternative optimal solutions is not easy until the model is solved. Once alternative optimal solutions are found, the user should determine which optimal solution is the most desirable in that situation.

Finding an optimal solution to an LP model can be regarded as assigning values to the decision variables so that the specified objective is achieved and the constraints are not violated. In cases in which an LP model consists of two decision variables only, it can be solved optimally by the graphical method. However, if an LP model possesses three or more decision variables, the simplex method or the interior point algorithm should be adopted. In the sections that follow, the well-known algorithms and methods for solving the LP models to optimality are discussed.

1.1.1 Simplex Method

The *simplex method*, introduced by G. B. Dantzig, has proved highly efficient in practice and therefore was widely adopted in commercial optimization packages for solving any LP model (Jensen and Bard 2003). Its development was based on the graphical method, which states that the optimal solution is always associated with a corner point of the solution space. The idea of the simplex method is to move the solution to a new corner that has the potential to improve the value of the objective function in each iteration. The process terminates when the optimal solution is found (Taha 2003).

Before applying the method, an LP must be converted into a standard form. The conditions of the standard form are that all constraints must be transformed into equality constraints and that all variables must be non-negative. If the constraint of an LP is a less-than-or-equal-to constraint, it can be converted into an equality constraint by adding a slack variable. If it is a greater-than-or-equal-to constraint, a surplus variable should be subtracted from the original constraint to become an equality constraint. A standard LP form aims at finding the basic solutions of the simultaneous linear equations. These basic solutions are exactly the corner point solutions of the solution space. The simplex method is then executed iteratively to search for the optimum from among these basic solutions.

The formal iterative steps of the simplex method are listed as (Winston and Venkataramanan 2003):

- Step 1: Obtain a basic feasible solution from the standard form.
- Step 2: Determine whether the current basic feasible solution is optimal.
- Step 3: If the current basic feasible solution is not optimal, then determine which nonbasic variable should become a basic variable and which basic variable should become a nonbasic variable to find a new basic feasible solution with a better objective function value.
- Step 4: Use elementary row operations to find the new basic feasible solution with the better objective function value. Return to Step 2.

1.1.2 Revised Simplex Method

The simplex method, mentioned in Section 1.1.1, can be referred to as the *tableau simplex method*. It searches for the optimal solution by moving from one feasible basis to another feasible basis, which hopefully leads to a better objective function value, based on the elementary row operations. There is a drawback to the tableau simplex method. Each tableau is generated from the tableau immediately preceding it, which may lead to computational round-off error. To minimize the adverse effect of computational round-off error, the revised simplex method is based on matrix algebra instead of elementary row operations. The round-off error in any tableau can be controlled by controlling the accuracy of inverse feasible basis. Once the inverse feasible basis is known, the entire simplex tableau can be computed (Taha 2003).

1.1.3 Dantzig–Wolfe Decomposition Algorithm

The *Dantzig–Wolfe decomposition algorithm* aims at improving computational efficiency. This was particularly useful when the memory and speed of computers were modest. In an LP model, the constraints may be decomposed into two different sets, including common or central constraints (i.e., constraints involving any variables or activities) and independent constraints (i.e., constraints involving unique variables or activities). In the absence of common constraints, the subproblems can be solved efficiently and almost independently (Taha 2003).

1.1.4 Karmarkar Interior Point Algorithm

The simplex method searches for the optimal solution along the corner points of the solution space, whereas the *Karmarkar interior point algorithm* looks for the optimum through the interior of the feasible region (Jensen and Bard 2003). Theoretically, the number of iterations needed to yield the optimum can grow exponentially. Therefore, the simplex method is regarded as an exponential time algorithm, whereas the interior point algorithm is regarded as a polynomial time algorithm. The interior point algorithm has theoretical importance in that it provides a bound on the computational effort required to solve a problem that is a polynomial function of its size (Winston and Venkataramanan 2003). Undoubtedly, it is effective for extremely large LP models.

1.1.5 Duality

The *dual LP model* is another form of the original or primal LP model. The primal and dual problems have a strong relationship. If the primal model has an optimal solution, then the dual model will have an optimal solution, and vice versa. The optimal solution values of the primal and dual models are always the same. In cases in which the computational complexity differs in the primal and dual LP models, we can select and solve the easier model. There are four rules governing the primal–dual conversion (Taha 2003):

1. A dual variable is defined for each primal constraint (i.e., when the primal model has m constraints, the dual model will have m decision variables).

2. A dual constraint is defined for each primal variable (i.e., when the primal model has n decision variables, the dual model will have n constraints).

3. The constraint coefficients of a primal variable define the left-side coefficients of the dual constraint, and its objective function coefficients define the right-side coefficients.
4. The objective coefficients of the dual constraint equal those on the right side of the primal constraints.

1.1.6 Sensitivity Analysis

Sensitivity analysis, one of the most important issues in LP, examines how the optimal solution of an LP model changes when the parameters (e.g., objective function coefficients and right-side values) of the model are varied. Generally, there are four types of changes in an LP's parameter: (1) changing the right-side values of the constraints, (2) adding a new constraint, (3) changing the objective function coefficients, and (4) adding a new variable or activity. The first two changes affect the feasibility of the current optimal solution, whereas the last two changes affect the optimality of the current solution (Taha 2003). Comparatively, changing the right-side values of the constraints and changing the objective function coefficients have attracted more attention than adding a new constraint and adding a new variable or activity.

1.2 Integer Linear Programming

Integer linear programming, or *integer programming (IP),* has been widely adopted as a method of modeling because some variables are not continuous but are integers in many cases in real life. Actually, IP is a subset of LP, with an additional constraint that some or all decision variables are restricted to integral values, depending on the type of IP. The general maximization-type IP model can be formulated as shown in Model 1.2.1.

Model 1.2.1 Standard maximization-type integer linear programming model

$$\text{Maximize } z = \sum_{j=1}^{n} c_j x_j \tag{1.2.1}$$

subject to

$$\sum_{j=1}^{n} a_{ij} x_i \leq b_i \quad \text{for all } i \tag{1.2.2}$$

All $x_j \geq 0$ and integer

The general minimization-type IP model can be formulated as shown in Model 1.2.2.

Model 1.2.2 Standard minimization-type integer linear programming model

$$\text{Minimize } z = \sum_{j=1}^{n} c_j x_j \tag{1.2.3}$$

subject to

$$\sum_{j=1}^{n} a_{ij} x_j \leq b_i \quad \text{for all } i \tag{1.2.4}$$

$$\sum_{j=1}^{n} a_{ij} x_j = b_i \quad \text{for all } i \tag{1.2.5}$$

All $x_j \geq 0$ and x_1 integer

Models 1.2.1 and 1.2.2 are almost the same as Models 1.1.1 and 1.1.2, respectively, except that there are integrality requirements in Models 1.2.1 and 1.2.2. Generally, there are three types of IP:

1. Pure integer linear programming is used if all variables must be integral, as is the case with Model 1.2.1.
2. Mixed integer linear programming (MILP) is used if only some of the variables must be integers, as is the case with Model 1.2.2.
3. Binary integer linear programming is used if all the variables must be either 0 or 1.

Unlike LP with the simplex method, a good IP algorithm for a very wide class of IP problems has not been developed (Williams 1999). Different algorithms are good with different types of problem. Generally, IP algorithms are based on exploiting the tremendous computational success of LP. Thus, before applying an IP algorithm, the integer restriction on the problem should be relaxed first to form an LP model. Starting from the continuous optimum point obtained from the LP model, integer constraints are incorporated repeatedly to modify the LP solution space in a manner that will eventually render the optimum extreme point, satisfying the integer requirements.

1.2.1 Branch-and-Bound Algorithm

In practice, the *branch-and-bound (B&B) algorithm* is widely used for solving IP models, especially MILP models (Williams 1999). The idea of the B&B

algorithm is to perform the enumeration efficiently so that not all combinations of decision variables must be examined. Sometimes, the terms *implicit enumeration, tree search,* and *strategic partitioning* are used, depending on the implementation of the algorithm (Jensen and Bard 2003).

The B&B algorithm starts with solving an IP model as an LP model by relaxing the integrality conditions. In cases in which the resultant LP solution or the continuous optimum is an integer, this solution will also be the integer optimum. Otherwise, the B&B algorithm sets up lower and upper bounds for the optimal solution. The branching strategy repetitively decreases the upper bound and increases the lower bound. The process terminates, provided that the processing list is empty (Castillo et al. 2002).

1.2.2 Cutting Plane Algorithm

As is the case with the B&B algorithm for solving IP models, the *cutting plane algorithm* relaxes the integrality requirements of the IP models and solves the resultant LP. But rather than repetitively imposing restrictions on the fractional variables, as is done in the B&B algorithm, extra constraints (i.e., cutting planes) are systematically added to the model, and the model is then resolved. The new solution to the further constrained model may or may not be an integer. By continuing the process until an integer solution is found or the model is shown to be infeasible, the IP model can be solved (Williams 1999; Jensen and Bard 2003).

1.3 Goal Programming

Goal programming (GP), invented by Charnes and Cooper (1961), is very similar to the LP model except that multiple goals are considered at the same time. Deviation variables (i.e., $d_1^+, d_1^-, d_2^+, d_2^-, ..., d_n^+, d_n^-$) are included in each goal equation to represent the possible deviations from goals. Deviation variables with positive signs refer to *overachievement,* which means that deviations are greater than the target value; those with negative signs indicate *underachievement,* which means that deviations are less than the target value. The objective function of a GP is to minimize deviations from desired goals. For each goal, there are three possible alternatives of incorporating deviation variables in the objective function. If both overachievement and underachievement of a goal are not desirable, then both d_i^+ and d_i^- are included in the objective function. If overachievement of a goal is regarded as unsatisfactory, then only d_i^+ is included in the objective function. If underachievement of a goal is regarded as unsatisfactory, then only d_i^- is included in the objective function. The general GP model in the form of MILP can be formulated as shown in Model 1.3.1.

Model 1.3.1 Standard goal programming model

$$\text{Minimize } z = \sum_i \left(d_i^+ + d_i^- \right) \qquad (1.3.1)$$

subject to

$$\sum_j a_{ij} x_j \le b_i \quad \text{for all } i \qquad (1.3.2)$$

$$\sum_j a_{ij} x_j - d_i^+ + d_i^- = b_i \quad \text{for all } i \qquad (1.3.3)$$

$$\text{All } x_j = 0 \text{ or } 1; \ d_i^+ \text{ and } d_i^- \ge 0$$

In this GP model, a_{ij} is the coefficient, whereas b_i is the right-side value. d_i^+ and d_i^- are overachievement and underachievement of goal i, respectively. The decision variable of the GP model is denoted as x_j. Objective function 1.3.1 minimizes the total deviations from the goals, while subject to system constraint set 1.3.2 and resource constraint set 1.3.3. Because all the objective function and constraint sets are in the linear form, it belongs to the LP type. In addition, decision variables (i.e., x_j) are binary, and deviation variables (i.e., d_i^+ and d_i^-) are continuous. Therefore, it is regarded as the mixed IP model.

In the next two sections, two algorithms for solving GP models are discussed, the weights method and the preemptive method. The common point of both methods is that they convert multiple goals into a single objective function.

1.3.1 Weights Method

In the *weights method*, positive weights are assigned to the goals of the problem. The weights, representing the relative importance of the goals, are determined subjectively by the user. For example, $w_1 = 1$ and $w_2 = 2$ implies that the second goal is twice as important as the first goal. After assigning the weights, the method converts the multiple goals into a single objective function by summing up the weighted goals.

1.3.2 Preemptive Method

In the *preemptive method*, priority levels (i.e., $P_1, P_2, ..., P_n$), representing the relative importance of the goals, are identified by the user. Goals with priority level P_1 are most important, followed by those with priority level P_2, and so on (i.e., $P_1 > P_2 > ... > P_n$). The preemptive method considers one goal at a time. Those with a higher priority level are considered first. Once they have been satisfied that there can be no further improvement, the next most important goals are then considered. The solution procedure is terminated if the solution obtained from a lower priority goal degrades any higher priority solutions.

1.4 SAS for Operational Research

SAS (statistical analysis system) has very comprehensive products for all aspects of operational research, including data analysis, optimization, and matrix algebra among many others, so it is beyond the scope of this book to discuss every aspect of SAS. We will just examine a small part that will be most useful to operational researchers. Most operational research tools for modeling, analysis, and problem solving are found in SAS/OR® (an operational research software developed by SAS), but some optimization features are also present in SAS/STAT (statistical analysis software), SAS/IML (interactive matrix programming with integration to R), SAS Enterprise Miner, and SAS/ETS (econometric and time series analysis software). A range of other products can be found at www.sas.com.

These days, there is plenty of software that can be used for operational research. The major reason why we selected SAS is because it has various optimization tools that can be used in a wide range of problems in operational research. Besides, SAS has strong data management capabilities that can handle very large datasets efficiently, and it can work with multiple datasets simultaneously. SAS has also a wide variety of statistical procedures and data-mining techniques.

Here is a list of some related SAS tools, but full range of SAS products can be obtained from SAS®:

- Base SAS: Data management and basic procedures
- SAS/STAT: Statistical analysis
- SAS/OR®: Operational research
- SAS/ETS: Econometrics and time series analysis
- SAS/IML: Interactive matrix language
- SAS/IRP: Inventory optimization
- SAS/SQL: Structural query language
- SAS/Enterprise Miner: Data mining with SAS

The aim of this book is to explore the optimization tools in SAS/OR®; it is assumed that the reader is familiar with the SAS base programming.

SAS programs comprise a set of statements to create SAS datasets and to run predefined procedures or other routines to get the results. There are also facilities to develop macros in SAS and some optional statements that usually appear on the top of the program. The common options that we use in this book are:

```
options linesize = 80;
options pagesize = 60;
options nonumber;
options nodate;
```

The two main sections of each program are DATA step and PROC step:

- The DATA steps allow the user to read one or more raw SAS datasets and to manipulate data.
- The PROC steps perform analysis on the data and produce results based on the procedure used.

The DATA step could be used before or after the PROC steps. The DATA step is used to prepare the data for use by one or more of the SAS procedure steps or to perform manipulation on the datasets obtained in the PROC step.

Often the external data need to be given to SAS for further analysis. SAS has a powerful procedure to import data from other packages, including text delimited or Excel files. The main procedure for importing data is PROC IMPORT. Similarly, PROC EXPORT can be used to export datasets from SAS to other format, such as text delimited or Excel files. In this book, we use this procedure repeatedly, hence it is recommended that the user reads more about this procedure. A full syntax of these procedures is given in Appendix 1.

In this book, we explain some advanced procedures because we assume that the reader is familiar with SASBASE.

The reader should note that SAS is not case sensitive. Throughout this book, we repeatedly use variables that are surrounded by "_" (e.g., _under-score_); these are specifically used as parameters in SAS. In all cases, these variables must be used without any change. We also use variables that start with "_" (e.g., _underscore); these variables are specifically used as parameters to the programs in this book. The user can assign any appropriate values to these variables.

In the next section, some features of SAS/OR® are explained.

1.4.1 SAS/OR®

SAS/OR® software includes a completely new generation of optimization procedures that make it easier to tackle a broad range of optimization problems. SAS/OR® comprises many of the analytical modeling and solution methods that are referred to collectively as operational research. Major areas of operational research work addressed by SAS/OR® include:

- Mathematical optimization, including constraints and noncon-straints optimization
- Project and resource scheduling, including critical path and resource-constrained project scheduling with a full range of facility for project management
- Discrete event simulation to enable systems to be studied in a modeling environment in which the long-term effects of alternative configurations and policies can be measured, statistically analyzed, and compared

TABLE 1.1

SAS/OR® Procedures

Procedure	Description
PROC NETFLOW	Network flow optimization for problems involving flows between nodes (locations) along arcs (node-to-node connections); solves shortest path, minimum cost flow, and maximum flow problems
PROC INTPOINT	Linear optimization using an interior point algorithm
PROC OPTQP	A solver for quadratic optimization; interior point problems
PROC OPTMILP	A solver for mixed integer linear optimization; branch-and-bound
PROC OPTMODEL	Optimization modeling language and access to all new optimization solvers (e.g., linear, general nonlinear, quadratic, mixed integer linear)
PROC OPTLP	Linear optimization; primal and dual simplex solvers, interior point solver
PROC CPM	A solver for planning, controlling, and monitoring a project
PROC PM	An advanced solver for project management problems
PROC GANTT	A procedure to produce a Gantt chart that is a graphical scheduling tool for the planning and control of a project
PROC DTREE	An interactive procedure for decision analysis
PROC GA	A procedure for genetic algorithms and general optimization
PROC BOM	A procedure to perform bill-of-material processing problems

The following procedures are still supported by SAS for legacy users, but it is suggested that PROC OPTMODEL is used instead (PROC QP will not even run in SAS 9.2).

PROC LP	A solver for linear, integer, mixed integer, and binary variable optimization based on the simplex method
PROC ASSIGN	A solver for assignment problems in which one set of items must be assigned to another set at the lowest total cost
PROC TRANS	A solver for transportation problems in which items must be moved from a set of supply locations to a set of demand locations at the lowest cost
PROC QP	A solver for quadratic problems
PROC NLP	A solver for general nonlinear programming problems, quadratic programming problems, and least-squares problems

The main part of SAS/OR® is optimization, which includes the procedures noted in Table 1.1.

1.4.2 Example of Using PROC OPTMODEL

The PROC OPTMODEL procedure includes the powerful modeling language and state-of-the-art solvers for various classes of optimization problems.

PROC OPTMODEL modeling language offers a modeling environment for building, solving, and maintaining mathematical programming models. PROC OPTMODEL provides an efficient environment for converting the

TABLE 1.2

A Transportation Problem With Four Warehouses in Different Cities

Warehouse	Pub 1	Pub 2	Pub 3	Pub 4	Pub 5	Pub 6	Pub 7	Pub 8	Supply
W1	10	15	12	13	15	10	20	15	1000
W2	8	12	15	10	12	16	12	17	1500
W3	12	12	16	14	12	15	10	12	2000
W4	20	10	20	12	15	14	17	12	2000
Demand	200	500	700	800	900	900	1000	1500	–

symbolic formulation of an optimization model into SAS. PROC OPTMODEL also simplifies data transformation to populate optimization models with data from SAS datasets.

PROC OPTMODEL can be used to build and solve optimization models, as well as to provide an environment for modeling tools. The results of optimization models built with PROC OPTMODEL can be saved in SAS datasets that may be submitted to other SAS product, including SASBASE and other optimization procedures in SAS/OR®.

1.4.2.1 An Introductory Example

As an introductory example, consider the following transportation programming:

Suppose there are four warehouses in different cities. They have 1000, 1500, 2000, and 2000 tons of paper accordingly. There are eight publishers in different locations. They ordered 200, 500, 700, 800, 900, 900, 1000, and 1500 tons of paper to publish some new books. The cost of delivering 1 ton of paper from each warehouse to publisher is listed in Table 1.2.

Assume that the data are saved in two SAS files as follows (see program "sasor_1_1.sas"):

```
* Program 1.1: An example of PROC OPTMODEL, populating data
(preparing SAS datasets prior to call OPTMODEL);

* Creating dataset of cost of shipments from each warehouse
to publisher;

data d_trans;
input warehouse pub1-pub8 supply;
datalines;
1    10   15   12   13   15   10   20   15   1000
2     8   12   15   10   12   16   12   17   1500
3    12   12   16   14   12   15   10   12   2000
4    20   10   20   12   15   14   17   12   2000
;
```

```
* Creating dataset of demands by each publisher;
data d_demand;
input pub demand;
datalines;
1 200
2 500
3 700
4 800
5 900
6 900
7 1000
8 1500
;
```

This code shows how the PROC OPTMODEL can be used to find the minimum cost for the transportation problem. This procedure takes the model as defined in the two SAS datasets—"d_trans" and "d_demand"—and finds the minimum cost flow (see program "sasor_1_2.sas").

Program 1.2 An Example of PROC OPTMODEL, Populating Data, Constructing Linear Programming, and Solving the Model.

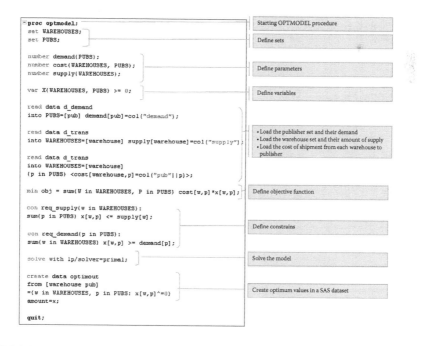

PROC OPTMODEL produces the following on the SAS log:

According to this, the minimum cost is $72,700. The solution of the PROC OPTMODEL is saved in the "optimout" dataset as follows:

	warehouse	pub	amount
1	1	3	100
2	1	6	900
3	2	1	200
4	2	3	500
5	2	4	800
6	3	3	100
7	3	5	900
8	3	7	1000
9	4	2	500
10	4	8	1500

The solution file shows the minimum cost of delivering papers from each warehouse to publishers.

1.4.2.2 Basic PROC OPTMODEL

PROC OPTMODEL is very powerful, so we can easily declare variables and parameters, define objective and constraints, and solve the problem. It also provides a full environment for programming using do-loop, if-then-else, and many other programming statements. The syntax of PROC OPTMODEL is given in Appendix 2. Here we give some details defining a linear programming within PROC OPTMODEL.

In most cases of defining a linear programming, we need to use the following six statements:

1. `number`: For defining confidences
2. `var`: For defining variables
3. `read`: For loading data from a dataset to the corresponding parameter

4. `min/max`: For defining an objective function
5. `con`: For defining a constraint
6. `solve`: For solving the problem using selected solver

Because in most linear programming we have a vector of variables and a matrix of coefficients, PROC OPTMODEL provides an indexing facility to handle these more efficiently. The index can be defined using integer numbers or a set of values. For example,

```
number c{1..4};
var x{1..4};
```

defines four numbers that can be referred to as c[1], c[2], c[3], and c[4] and defines four variables that can be referred to as x[1], x[2], x[3], and x[4].

Using the index provides an easy environment for working with parameters. For example, the following statement finds the sum of the four parameters just mentioned:

```
S = sum{i in 1..4} x[i];
```

The following statements define an objective function in the form of $\min \sum_{j=1}^{4} c_j x_j$:

```
min z = sum{j in 1..4}(c[j]*x[j]);
```

The following code constructs (see program " sasor_1_3.sas") and solves a model for the following linear programming:

$$\text{Max } 4x_1 + 2x_2 + x_3 + 3x_4$$

subject to

$$2x_1 + x_2 + 4x_4 \leq 4$$

$$4x_1 - 1x_2 + 4x_3 \leq 3$$

$$3x_1 + 2x_2 + x_3 + 2x_4 \leq 8$$

$$x_1, x_2, x_3, x_4 \geq 0$$

```
*Program 1.3: An example of PROC OPTMODEL;
proc optmodel;

* Declare parameters a, b and c;
number c{1..4} = [4, 2, 1, 3] ;
number b{1..3} = [4, 3, 8];
number a{1..3, 1..4} = [2, 1, 0, 4,
                        4, -1, 4, 0,
                        3, 2, 1, 2];
print a;

* Declare variable x;
var x{1..4} > = 0;

* Define objective function;
max z = sum{i in 1..4}(c[i]*x[i]);

* Define constraints;
con constraint{i in 1..3}:
sum{j in 1..4} a[i,j]*x[j] < = b[i];

* Print the linear programming;
expand;

* Solve the model;
solve;

* Print optimum solution for x;
print x.sol;
%put &_OROPTMODEL_;
quit;
```

PROC OPTMODEL produces the following on the SAS log showing that there are four variables and three linear constraints and that the objective value is 8.7.

```
Log - (Untitled)                                                              _ □ ×
NOTE: The problem has 4 variables (0 free, 0 fixed).
NOTE: The problem has 3 linear constraints (3 LE, 0 EQ, 0 GE, 0 range).
NOTE: The problem has 10 linear constraint coefficients.
NOTE: The problem has 0 nonlinear constraints (0 LE, 0 EQ, 0 GE, 0 range).
NOTE: The OPTLP presolver value AUTOMATIC is applied.
NOTE: The OPTLP presolver removed 0 variables and 0 constraints.
NOTE: The OPTLP presolver removed 0 constraint coefficients.
NOTE: The presolved problem has 4 variables, 3 constraints, and 10 constraint coefficients.
NOTE: The DUAL SIMPLEX solver is called.
NOTE:                      Objective
      Phase Iteration  Value
         2       1        11.333333
         2       3         8.700000
NOTE: Optimal.
NOTE: Objective = 8.7.
```

The expanded statement in the Program 1.3 code shows the linear programming model in the output window as follows:

```
VAR X[1] > = 0
VAR X[2] > = 0
VAR X[3] > = 0
VAR X[4] > = 0
Maximize z = 4*x[1] + 2*x[2] + x[3] + 3*x[4]
Constraint constraint[1]: 2*x[1] + x[2] + 4*x[4] < = 4
Constraint constraint[2]: 4*x[1] − x[2] + 4*x[3] < = 3
Constraint constraint[3]: 3*x[1] + [2]*x[2] + x[3] + 2*x[4] < = 8
```

The "print x.sol" shows the optimum solution of the x variables in the output window, which is:

The OPTMODEL Procedure

Solution Summary

Solver	Dual Simplex
Objective function	z
Solution status	Optimal
Objective value	8.7
Iterations	3
Primal infeasibility	4.163336E-17
Dual infeasibility	0
Bound infeasibility	0

[1]	x.SOL
1	0.7
2	2.6
3	0.7
4	0.0

1.4.2.3 Set and Indexing in PROC OPTMODEL

Generally, parameters and expressions can have numerical or character values. For example, the number statement used in the previous code declares a numerical variable, while with the set statement we can define both numerical and string variables. Consider the following codes that define two sets of rows and columns and initialize the bank data with the following matrix

$$\begin{bmatrix} \text{Labor} & \text{Capital} & \text{Profit} \end{bmatrix}$$

$$\begin{bmatrix} \text{Bank1} \\ \text{Bank2} \\ \text{Bank3} \\ \text{Bank4} \end{bmatrix} \begin{bmatrix} 10 & 2000 & 30 \\ 50 & 40000 & 68 \\ 8 & 25000 & 45 \\ 18 & 70000 & 50 \end{bmatrix}$$

```
*Program 1.4: Set and indexing in PROC OPTMODEL;

 proc optmodel;
   set < string > row ;
   set < string > col ;
   row = {"Bank1", "Bank2", "Bank3", "Bank4"};
   col = {" Labor", "Capital", "Profit"};

   number bank{r in row, c in col} =
           [10, 2000, 30,
            50, 40000, 68,
            8, 25000, 45,
            18, 70000, 50];
   print bank;
 quit;
```

Using this definition enables us to refer to the elements of the bank matrix using index variables. For example Bank["Bank2", "Labor"] equals 50 and Bank["Bank3", "Capital"] equals 25,000.

An alternative initialization of the data to parameters in PROC OPTMODEL is using the "read" statement and populating parameter with the data saved in a dataset. Assume that the data in Program 1.4 are saved in a bank dataset; the following program reads the dataset and loads it to the corresponding variables:

```
*Program 1.5: An example of PROC OPTMODEL using read
statement;

 data bankdata;
   input Bank $ Labor Capital Profit;
   datalines;
   Bank1 10 2000 30
   Bank2 50 40000 68
   Bank3 8 25000 45
   Bank4 18 70000 50
 ;
```

```
 proc optmodel;
   * Define parameters;
   set < string > row ;
   set < string > col ;
   col = {"Labor", "Capital", "Profit"};
   number bankmatrix{r in row, c in col};

   * Populating name of banks from the first column of the
   dataset to 'row';
   read data bankdata
   into row = [Bank];
```

```
* Populating value of Capital, Labor, and Profit to each
bank from the dataset;
read data bankdata
into
{r in row} < bankmatrix[r, " Labor"] = col(" Labor")
                bankmatrix[r, "Capital"] = col("Capital")
                bankmatrix[r, "Profit"] = col("Profit") > ;
* Printing bankmatrix;
print bankmatrix;

quit;
```

In this code, we used a "read" statement. The first "read" loads the bank names to set row while the second "read" loads the value of capital, labor, and profit to each bank.

1.4.2.4 Advanced Options in PROC OPTMODEL

As discussed earlier, PROC OPTMODEL provides a full environment for programming using do-loop, if-then-else, and many other programming statements. We can divide the syntax of PROC OPTMODEL into three types of statements:

1. Options in PROC OPTMODEL
2. Declaration of parameters and variables, as well as objective function and constraints
3. Programming statements

With the option statements, you can control how the optimization model is processed and how results are displayed. The declaration statements define the parameters, variables, constraints, and objectives that describe the model to be solved. All declarations in the PROC OPTMODEL are also saved for later use. The most popular declaration statements are:

- constraint (or con): Defines one or more constraints
- max/min: Declares an objective for the solver
- number (or num): Declares a numeric parameter
- string (or str): Declares a string parameter
- set: Declares a set type parameter
- var: Declares a variable

Parameters and variables can also be initialized using option "init."

In the programming statements of the PROC OPTMODEL, you can program the reading, calculating, and writing of the results. The five main categories of programming statements include:

1. Loop statements
   - Continue: Terminates the current iteration of the loop statement (iterative DO, DO UNTIL, DO WHILE, or FOR)
   - For: Executes the substatements for each member of the specified index set in the "for" statement
   - Do iterative: Defines an iterative "do" statement
   - Do until: Executes the specified sequence of statements repeatedly until the logic-expression, evaluated after the statements, returns true
   - Do while: Executes the specified sequence of statements repeatedly as long as the logic-expression, evaluated before the statements, returns a true nonzero value
   - Leave: Terminates the execution of the entire loop body (iterative DO, DO UNTIL, DO WHILE, or FOR) that immediately contains the LEAVE statement

2. Control statements
   - Do: Groups a sequence of statements together as a single statement; using "do; statements; end"
   - If-then-else: Evaluates the logical expression and then conditionally executes the THEN or ELSE substatements
   - Stop: Ends the execution of all statements that contain it, including "do" statements and other control or looping statements

3. General statements
   - Assignment: Assigns a value to a parameter
   - Call: Calls a subroutine that is defined in another part of the SAS code
   - Reset options: Restores option values of parameters to their default values

4. Input/output statements
   - Close file: Closes files that were opened by the "file" statement
   - Create data: Creates a new SAS dataset and copies data into it
   - File: Defines a file name that can be selected from the current output file for the PUT statement
   - Print: Prints string and numerical data in tabular form
   - Put: Writes text data to the current output file that is assigned in the "File" statement

- Read data: Reads data from a SAS dataset into parameter and variable locations
- Save MPS: Saves the structure and coefficients for the defined linear programming model into a SAS dataset in the MPS format, which later can be used as input data for the PROC OPTLP or PROC OPTMILP procedure
- Save QPS: Saves the structure and coefficients for the defined linear programming model into a SAS data set in the APS format, which later can be used as input to PROC OPTQP procedure

5. Model statements

- Drop: Ignores the specified constraint, constraint array, or constraint array location from the model
- Expand: Prints the specified constraint, variable, or objective declaration expressions after expanding aggregation operators, substituting the current value for parameters and indices, and resolving constant subexpressions
- Fix: Tells the solver to treat a list of variables, variable arrays, or variable array locations as fixed in value
- Restore: Adds a constraint, constraint array, or constraint array location that was dropped by the "drop" statement back into the solver model
- Solve: Solves the defined model
- Unfix: Reverses the effect of "fix" statements

The PROC OPTMODEL procedure produces a macro variable (_OROPTMODEL_) at termination. This variable contains a character string that indicates the status of the procedure on termination and gives the objective value and number of iteration to solve the problem. An example of this macro is:

```
STATUS = OK SOLUTION STATUS = OPTIMAL OBJECTIVE = 8.7 PRIMAL_INFEASIBILITY =
4.163336E-17 DUAL_ INFEASIBILITY = 0 BOUND_ INFEASIBILITY = 0 ITERATIONS = 3
PRESOLVE_TIME = 0.015 SOLUTION_TIME = 0 694 quit;
```

The syntax of PROC OPTMODEL is given in Appendix 2, and Chapter 3 covers several examples of PROC OPTMODEL.

Earlier we demonstrated how the PROC OPTMODEL procedure can be used for solving optimization problems. There are many other procedures available in SAS/OR® as shown in Table 1.1. The principles of the remaining procedures will be presented in the following sections. Some of them are briefly explained here and examples are given in Chapter 3. The reader who is interested in these procedures may refer to *SAS/OR® User's Guide: Mathematical Programming*.

1.4.3 Other Procedures in SAS/OR®*

1.4.3.1 PROC OPTLP

PROC OPTLP concentrates on the solution of linear programs, offering three solvers: primal simplex, dual simplex, and an experimental interior point algorithm. It includes an aggressive presolver that can work to reduce the effective size of the optimization model before the solver begins its work. The PROC OPTLP procedure accepts models specified in an SAS dataset that uses the MPS data format, which has become a standard in linear optimization.

1.4.3.2 PROC OPTMILP

PROC OPTMILP is a procedure for solving mixed integer linear programs using a B&B algorithm based on the simplex method. This technique involves the sequential creation and solution of a series of related linear programs, with new and modified linear programs potentially being generated at each step. The PROC OPTMILP procedure includes a presolver for reducing the effective size of the model and also employs cutting planes and primal heuristics to speed up the progress of the B&B algorithm.

1.4.3.3 PROC OPTQP

The PROC OPTQP procedure provides linear programming solvers and enables you to choose from three linear programming solvers: primal simplex, dual simplex, and interior point. Presolvers, which work aggressively to reduce the effective size of problems before the solvers are invoked, are also provided. The simplex solvers implement a two-phase simplex method, and the interior point solver implements a primal–dual predictor–corrector algorithm. This solver is especially designed to handle large-scale problems because many quadratic programming problems originate in large corporate settings and are broad in scope. The PROC OPTQP procedure handles both sparse and dense problems well. For the PROC OPTQP procedure, the problem should be defined in an SAS dataset adhering to the QPS data format, an extension of the MPS format.

1.4.3.4 PROC INTPOINT

The PROC INTPOINT procedure solves the network program with side constraints and the more general linear programming problem through use of the interior point algorithm. The data required by PROC INTPOINT are similar to the data required by PROC NETFLOW for solving network flow models using the interior point algorithm.

---

* Source: www.sas.com

1.4.3.5 PROC CPM

The PROC CPM procedure can be used for planning, controlling, and monitoring a project. A typical project consists of several tasks that may have precedence and time constraints. PROC CPM enables you to schedule activities subject to all other constraints.

PROC CPM enables you to define calendars and specify holidays for the different activities so that you can schedule around holidays and vacation periods. Once a project has started, you can monitor it by specifying current information or progress data that is used by PROC CPM to compute an updated schedule. You can compare the new schedule with a baseline (or target) schedule.

For projects with scarce resources, you can determine resource-constrained schedules. PROC CPM enables you to choose from a wide variety of options so that you can control the scheduling process. Thus, you may choose to allow project completion time to be delayed or use supplementary levels of resources, or alternate resources, if they are available.

1.4.3.6 PROC PM

PROC PM enhances the extensive project management capabilities of SAS/OR® software by providing a graphical user interface for creating and editing project activity data. PROC PM procedure's Project View combines a Table View and a Gantt View to enable you to define and modify your project models interactively, with immediate visual confirmation of the effects on the project schedule. Additional capabilities in PROC PM include being able to:

- Use the Gantt View to simultaneously examine up to four different versions of the project schedule.
- Edit activity data directly via the Table View or via drag-and-drop in the Gantt View.
- Add and delete activities, subactivities, projects, and subprojects.
- Establish and modify precedence relationship.
- Alter activity durations and time alignments.
- Reorder the Table View data columns.
- Hide activities or shuffle their order.
- Expand or collapse projects and subprojects to display just the required activities and data.

PROC PM is built on the foundation of PROC CPM's syntax, ensuring backward compatibility with PROC CPM. Thus PROC PM can be easily integrated into any existing application that is built around PROC CPM and the other established project management tools in the SAS System.

1.4.3.7 PROC GANTT

PROC GANTT procedure produces a Gantt chart that is a graphical scheduling tool for the planning and controlling of a project. In its most basic form, a Gantt chart is a bar chart that plots the tasks of a project versus time. PROC GANTT displays a Gantt chart corresponding to a project schedule, such as that produced by PROC CPM or one that is input directly to the procedure, and it offers several options and statements for tailoring the chart to your needs.

Using PROC GANTT, you can plot the predicted early and late schedules and identify critical, supercritical, and slack activities. In addition, you can visually monitor a project in progress with the actual schedule and compare the actual schedule against a target baseline schedule. You can also graphically view the effects of scheduling a project that is subject to resource limitations. Any combination of these schedules can be viewed simultaneously (provided that the relevant data exist) together with any user-specified variables of interest, such as project deadlines and other important dates. PROC GANTT enables you to display the early, late, and actual schedules in a single bar to produce a more meaningful schedule for tracking an activity in progress.

1.4.3.8 PROC DTREE

The PROC DTREE procedure in SAS/OR® software is an interactive procedure for decision analysis. The procedure interprets a decision problem represented in SAS datasets, finds the optimal decisions, and plots on a line printer or a graphical device the decision tree showing the optimal decisions.

To use PROC DTREE, you first construct a decision model to represent your problem. This model, called a *generic decision tree model*, is made up of stages. Every stage has a stage name, which identifies the stage, a type (which specifies the type of the stage), and the possible outcomes of the stage. There are three types of stages: decision stages, chance stages, and end stages.

1.4.3.9 PROC GA

The PROC GA procedure facilitates the application of genetic algorithms to general optimization. PROC GA is especially useful for finding the optimal solution for problems in which the objective function may have discontinuities or may not otherwise be suitable for optimization by traditional calculus-based methods.

1.4.3.10 PROC BOM

The PROC BOM procedure performs bill-of-material (BOM) processing. It is composed of a series of single-level BOMs divided into a multilevel, tree-structured BOM; determines the level of each part in the bill; and represents the multilevel BOM structure in the form of an indented BOM. PROC BOM can also output a summarized BOM.

A BOM is a list of all parts, ingredients, or materials needed to make one production run of a product. The BOM may also be called the formula, recipe, or ingredients list in certain process industries (Cox and Blackstone 1998). The way in which the BOM data are organized and presented is called the BOM structure or the product structure. The indented BOM data generated by PROC BOM are organized in such a manner that they can be easily retrieved and manipulated to generate reports and can also be used by other SAS/OR® procedures to perform additional analysis. The summarized BOM data are quite useful in gross requirements planning and other applications.

For the full syntax of these procedures, we refer the reader to the SAS/OR® manual; other related information about SAS products for operational researchers is listed in Appendix 3.

2

Transportation Models

In this chapter, we present the family of transportation models and demonstrate how SAS/OR® can be applied to solve transportation, assignment, and transshipment problems to optimality. The problem formulations are described first. Then, various SAS/OR® procedures are applied to tackle the problems with the aid of examples. Following that, result analyses are carried out. After this chapter, the reader will be more familiar with SAS/OR® and the applications of its procedures.

2.1 Transportation Problem

2.1.1 Concept of Transportation Problem

The *transportation problem*, first described by Hitchcock in 1941, is a special class of linear programming (LP) problem. The objective is to yield the least-cost means of shipment through a transportation network in which there is a set of origins providing a commodity to a definite number of destinations. Suppose that a number of suppliers ($i = 1, 2, ..., m$) provides a commodity to a number of customers ($j = 1, 2, ..., n$). The transportation problem determines to meet each customer's requirement, d_j, while not exceeding the capacity of any supplier, s_i, at minimum cost, c_{ij}. By introducing variables x_{ij} to represent the quantity of the commodity sent from supplier i to customer j, the transportation model can be written as shown in Model 2.1.1.

Model 2.1.1 Standard transportation model

$$\text{Minimize } z = \sum_{i=1}^{m} \sum_{j=1}^{n} c_{ij} x_{ij} \tag{2.1.1}$$

subject to

$$\sum_{j=1}^{n} x_{ij} \leq s_i \quad i = 1, 2, ..., m \tag{2.1.2}$$

$$\sum_{i=1}^{m} x_{ij} \geq d_j \quad j = 1, 2, \ldots, n \qquad (2.1.3)$$

$$\text{All } x_{ij} \geq 0$$

Model 2.1.1 is referred to as the *transportation model*. Objective function 2.1.1 minimizes the total transportation cost. Unit transportation costs c_{ij} for shipping 1 unit of commodity from supplier i to customer j are known. These costs are often dependent on the travel distances between supplier i to customer j. It is assumed that the cost on a particular route of the transportation network is directly proportional to the number of commodities shipped on that route. If supplier i cannot supply customer j, the unit transportation cost c_{ij} is considered infinite (∞). Constraint set 2.1.2 is known as a *supply constraint* or *availability constraint*, and constraint set 2.1.3 is known as a *demand constraint* or *requirement constraint*. It is assumed that the capacity of each supplier, s_i, and the demand of each customer, d_j, are known in advance. If the total supply equals the total demand, then the problem is said to be a balanced transportation problem. In this case, constraint sets 2.1.2, and 2.1.3 are treated as both equal instead of less than or equal to and greater than or equal to, respectively. If the total supply does not equal the total demand, then the problem is referred to as an *unbalanced transportation problem*. A dummy customer (when the total supply exceeds the total demand) or a dummy supplier (when the total demand exceeds the total supply) is added to balance the transportation model. Because shipments via the dummy supplier or dummy customer are not real shipments, the unit transportation costs assigned to them are 0.

2.1.2 Example of Transportation Problem

Figure 2.1 shows a transportation network in which there are four suppliers and five customers. The unit transportation costs are shown above the arcs, or arrows. For example, it costs 3 units of dollars to ship 1 unit of commodity from supplier 1 to customer 1. The capacity of each supplier, s_i, and the demand of each customer, d_j, are also shown.

This transportation network or problem can be represented by a tableau as shown in Table 2.1. The upper-right corner of each cell in the tableau represents the unit transportation cost c_{ij}.

By introducing variables x_{ij} to represent the quantity of the commodity shipped from supplier i to customer j, this transportation problem can be formulated as shown in Model 2.1.2.

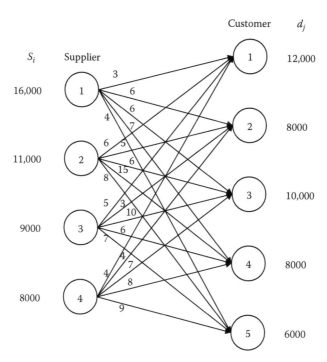

FIGURE 2.1
A transportation network.

TABLE 2.1

A Transportation Problem

Supplier i	Customer j					s_i
	1	**2**	**3**	**4**	**5**	
1	3 x_{11}	6 x_{12}	6 x_{13}	7 x_{14}	4 x_{15}	16000
2	6 x_{21}	5 x_{22}	6 x_{23}	15 x_{24}	8 x_{25}	11000
3	5 x_{31}	3 x_{32}	10 x_{33}	6 x_{34}	7 x_{35}	9000
4	4 x_{41}	4 x_{42}	7 x_{43}	8 x_{44}	9 x_{45}	8000
d_j	12000	8000	10000	8000	6000	

Model 2.1.2 Example of formulation of transportation problem

$$\text{Minimize} \quad 3\,x_{11} + 6\,x_{12} + 6\,x_{13} + 7\,x_{14} + 4\,x_{15}$$

$$+ 6\,x_{21} + 5\,x_{22} + 6\,x_{23} + 15\,x_{24} + 8\,x_{25}$$

$$+ 5\,x_{31} + 3\,x_{32} + 10\,x_{33} + 6\,x_{34} + 7\,x_{35}$$

$$+ 4\,x_{41} + 4\,x_{42} + 7\,x_{43} + 8\,x_{44} + 9\,x_{45} \qquad (2.1.4)$$

subject to

$$x_{11} + x_{12} + x_{13} + x_{14} + x_{15} = 16{,}000 \qquad (2.1.5)$$

$$x_{21} + x_{22} + x_{23} + x_{24} + x_{25} = 11{,}000 \qquad (2.1.6)$$

$$x_{31} + x_{32} + x_{33} + x_{34} + x_{35} = 9000 \qquad (2.1.7)$$

$$x_{41} + x_{42} + x_{43} + x_{44} + x_{45} = 8000 \qquad (2.1.8)$$

$$x_{11} + x_{21} + x_{31} + x_{41} = 12{,}000 \qquad (2.1.9)$$

$$x_{12} + x_{22} + x_{32} + x_{42} = 8000 \qquad (2.1.10)$$

$$x_{13} + x_{23} + x_{33} + x_{43} = 10{,}000 \qquad (2.1.11)$$

$$x_{14} + x_{24} + x_{34} + x_{44} = 8000 \qquad (2.1.12)$$

$$x_{15} + x_{25} + x_{35} + x_{45} = 6000 \qquad (2.1.13)$$

$$\text{All } x_{ij} \geq 0$$

Constraint sets 2.1.5 to 2.1.8 are the availability constraints. There is one such constraint for each supplier. For example, constraint set 2.1.5 ensures that the total amount of commodities shipped from supplier 1 equals the supplier's capacity. Constraint sets 2.1.9 to 2.1.13 are the requirement constraints. Similarly, there is one such constraint for each customer. For example, constraint set 2.1.9 ensures that the total amount of commodities ordered by customer 1 is fulfilled. Although Model 2.1.2 is an LP model, the optimal solution must be integral. This is a characteristic of the transportation problem.

2.1.3 ORTRANS: SAS Code for Transportation Problem

ORTRANS is a macro that solves transportation problems, the objective of which is to yield the minimum cost of shipment through a transportation

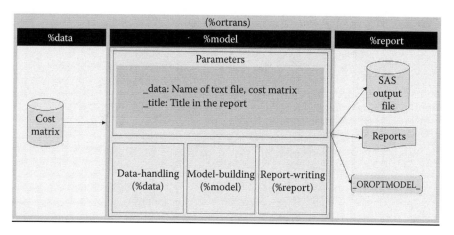

FIGURE 2.2
Data flow in ORTRANS.

network to given destinations (see program "sasor_2_1.sas"). The main procedure used for such a transportation problem is PROC OPTMODEL.

Figure 2.2 illustrates the data flow in the ORTRANS. It shows:

- The cost matrix that is required for ORTRANS, in which the capacity of any supplier (s_i) and the customer requirement (d_j) are specified
- The macros (%data, %model, and %report)
- The result data sets that are available for print or can be used for further analysis

In the rest of this section, the procedure for solving a transportation problem (ORTRANS) in SAS, together with an example, is explained. ORTRANS runs three macros: data-handling (%data), model-building (%model), and report-writing (%report).

2.1.4 ORTRANS Data-Handling Macro (%data)

This part of ORTRANS processes the data into a format that is suitable for PROC OPTMODEL. ORTRANS requires one data set containing names the of suppliers and customers and the cost matrix. The data set should be a .txt file, which is saved as "text tab delimited." The suppliers' and customers' names must start with a letter and may contain up to 50 characters. The customers' names must be listed in the first row of the data file. The amount of supply and suppliers' name should be listed in columns named "supply" and "supplier," respectively. An example of a data file is seen in Figure 2.3.

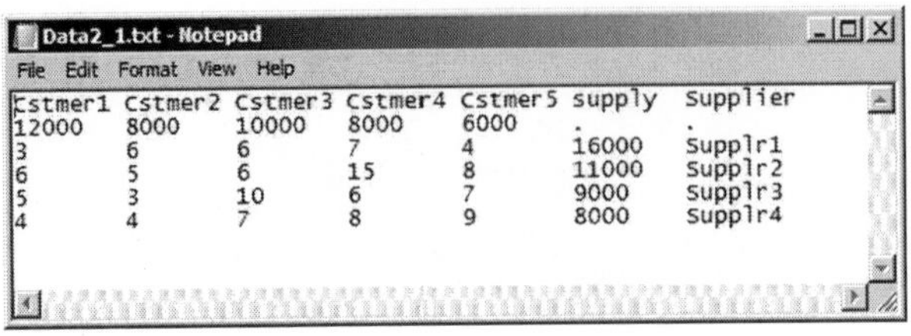

Cstmer1	Cstmer2	Cstmer3	Cstmer4	Cstmer5	supply	Supplier
12000	8000	10000	8000	6000	.	.
3	6	6	7	4	16000	Supplr1
6	5	6	15	8	11000	Supplr2
5	3	10	6	7	9000	Supplr3
4	4	7	8	9	8000	Supplr4

FIGURE 2.3
An example of a dataset, cost matrix.

One parameter needs to be set before calling the data macro:

_data: Indicates the name and location of the data file (a text tab delimited file) and contains cost matrix

```
* The data-handling macro;
%macro data;

* Import text tab delimited data file to SAS data file;
  proc import
    datafile = &_data
    out = dcost
    dbms = tab
    replace;
    getnames = yes;
  run;

%mend data;
```

2.1.5 ORTRANS: Model-Building Macro (%model)

This part of ORTRANS calls PROC OPTMODEL to solve the model. The SAS macro for model-building is as follows:

```
* Starting OPTMODEL Procedure;
proc optmodel;

* Define sets;
set < string > SUPPLIERS;
set CUSTOMERS = {1..5};

* Define parameters;
number demand{CUSTOMERS};
number supply{SUPPLIERS};
number cost{SUPPLIERS, CUSTOMERS};

* Define variables;
var X{SUPPLIERS, CUSTOMERS} > = 0;
```

```
* Load the supplier set and their amount of supply;
read data dcost (where = (supply ne .))
into SUPPLIERS = [Supplier] supply[Supplier] =
col ("supply");

* Load the customer set and their demand;
read data dcost (where = (supply eq .)) into
{c in CUSTOMERS} < demand[c] = col("Cstmer"||c) > ;

* Load the cost of shipment from each supplier to each
customer;
read data dcost (where = (supply ne .))
into SUPPLIERS = [Supplier]
{c in CUSTOMERS} < cost[Supplier,c] = col("Cstmer"||c) > ;

* Define objective function;
min obj = sum{s in SUPPLIERS, c in CUSTOMERS}
cost[s,c]*x[s,c];

* Define constraints;
con req_supply{s in SUPPLIERS}:
sum{c in CUSTOMERS} x[s,c] < = supply[s];
con req_demand{c in CUSTOMERS}:
sum{s in SUPPLIERS} x[s,c] > = demand[c];

con zero{c in CUSTOMERS, s in SUPPLIERS : cost[s,c] =
1E10}: x[s,c] = 0;

* Solve the model;
solve with lp/solver = primal;

* Create optimum values in an SAS dataset 'optimout';
create data optimout
from [SUPPLIERS CUSTOMERS]
= {s in SUPPLIERS, c in CUSTOMERS: x[s,c]^ = 0}
amount = x;

* End of OPTMODEL Procedure;
quit;
```

2.1.6 ORTRANS: Report-Writing Macro (%report)

The output from ORTRANS includes results and a table of cost, which is saved in an SAS DATA file named "result" and "dcost", respectively, as well as in print format. The SAS procedure for report-writing is as follows. Two further datasets, "result1" and "result2", show the information sorted by suppliers' names and the amount of supply, respectively. One parameter needs to be set before calling this macro:

_title: Gives a title in the output of the SAS

```
* The report-writing macro;
%macro report;
 title &_title;
 * Sort results by supplier;
 * report the results in a tabulated form;
proc tabulate data = optimout;
title &_title;
class SUPPLIERS CUSTOMERS ;
var amount;
table SUPPLIERS = " Suppliers",
     CUSTOMERS*amount*sum
     / BOX = 'Amount of suppliers to customers' ;
run;

%mend report;
```

2.1.7 ORTRANS: Macro (%ortrans)

To make the system as user friendly as possible, the %ortrans macro combines the data-handling, model-building, and report-writing codes.

```
* The ortrans macro for transportation problem;
%macro ortrans;
 %data;
 %model;
 %report;
%mend ortrans;
```

In this code, the %ortrans macro is used to manage all the codes explained earlier, including data-handling, model-building, and report-writing. To get the result, the user needs to set up the parameters and run only one statement:

```
%ortrans;
```

2.1.8 Instructions for Using ORTRANS Macro

This section presents SAS code for the earlier transportation problem with four suppliers and five customers as shown in Table 2.1. The data are saved in file "data2_1.txt."

A user needs to set the parameters as required and run the following code:

```
* SAS macro for transportation problem;
%let _title = 'Example 2.1: Transportation problem for four
suppliers and five customers.';
%let _data = 'c:\sasor\data2_1.txt';
%ortrans;
```

Amount of suppliers to customers	CUSTOMERS				
	1	2	3	4	5
	Amount	Amount	Amount	Amount	Amount
	Sum	Sum	Sum	Sum	Sum
Suppliers					
Supplr1	10000.00	.	.	.	6000.00
Supplr2	.	1000.00	10000.00	.	.
Supplr3	.	1000.00	.	8000.00	.
Supplr4	2000.00	6000.00	.	.	.

FIGURE 2.4
Result of %ortrans sorted by suppliers' name.

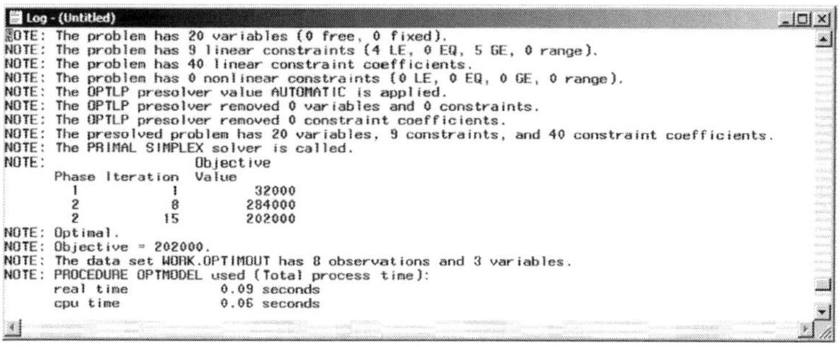

```
Log - (Untitled)
NOTE: The problem has 20 variables (0 free, 0 fixed).
NOTE: The problem has 9 linear constraints (4 LE, 0 EQ, 5 GE, 0 range).
NOTE: The problem has 40 linear constraint coefficients.
NOTE: The problem has 0 nonlinear constraints (0 LE, 0 EQ, 0 GE, 0 range).
NOTE: The OPTLP presolver value AUTOMATIC is applied.
NOTE: The OPTLP presolver removed 0 variables and 0 constraints.
NOTE: The OPTLP presolver removed 0 constraint coefficients.
NOTE: The presolved problem has 20 variables, 9 constraints, and 40 constraint coefficients.
NOTE: The PRIMAL SIMPLEX solver is called.
NOTE:                     Objective
      Phase Iteration     Value
        1          1          32000
        2          8         284000
        2         15         202000
NOTE: Optimal.
NOTE: Objective = 202000.
NOTE: The data set WORK.OPTIMOUT has 8 observations and 3 variables.
NOTE: PROCEDURE OPTMODEL used (Total process time):
      real time           0.09 seconds
      cpu time            0.06 seconds
```

FIGURE 2.5
Log for %ortrans.

This code manages to get the results based on the specified parameters and the cost matrix saved in the text file. At termination, this code also produces a macro variable (_OROPTMODEL_) containing a character string that indicates the status of the procedure on termination and provides the objective value at termination. Because _OROPTMODEL_ is a standard SAS macro variable, it can be used in the ways that all macro variables can be used. See the *SAS Guide to Macro Processing* for more information.

2.1.9 Sample Results from ORTRANS Macro: Output from SAS

The results of running the ORTRANS macro code are presented in Figure 2.4, in which the solutions are sorted by suppliers' name and amount of supply. According to the results, 10,000 units of products are shipped from supplier 1 to customer 1 (i.e., $x_{11} = 10{,}000$), 6000 units of products are shipped from supplier 1 to customer 5 (i.e., $x_{15} = 6000$), and so on. The optimal total transportation cost is \$202,000, which is shown in Figure 2.5. The total computational time is 0.06 second.

The solution to the transportation problem can also be obtained and reported in a tabular format, as shown in Figure 2.4.

The log file in Figure 2.5 shows the number of variables and constraints, and it also shows how we get the objective value of $202,000 using the %ortrans macro.

2.1.10 Exercise

Use the codes developed in this section and solve the transportation problem in Table 2.2.

Solution:

- Create the data in a text file (see "data2_1_exercise.txt").
- Run the following code (see program " sasor_2_1_exercise.sas").

```
* SAS macro for transportation problem: solution to
exercise 2.1;
%let _title = ' Transportation problem: solution to
exercise 2.1.';
%let _data = 'c:\sasor\data2_1_exercise.txt';
%ortrans;
```

The following solution is given by SAS:

Amount of suppliers to customers	CUSTOMERS			
	1	2	3	4
	Amount	Amount	Amount	Amount
	Sum	Sum	Sum	Sum
Suppliers				
Supplr1	10.00	.	80.00	.
Supplr2	.	80.00	40.00	.
Supplr3	90.00	.	.	60.00

TABLE 2.2

A Transportation Exercise

Supplier i	Customer j				s_i
	1	2	3	4	
1	4 x_{11}	7 x_{12}	7 x_{13}	1 x_{14}	90
2	12 x_{21}	3 x_{22}	8 x_{23}	8 x_{24}	120
3	8 x_{31}	10 x_{32}	16 x_{33}	5 x_{34}	150
d_j	100	80	120	60	

```
Log - (Untitled)                                                        _ □ ×
NOTE: The presolved problem has 12 variables, 7 constraints, and 24 constraint coefficients.
NOTE: The PRIMAL SIMPLEX solver is called.
NOTE:                    Objective
     Phase Iteration    Value
         1        1        270.000000
         2        6       2580.000000
         2        8       2180.000000
NOTE: Optimal.
NOTE: Objective = 2180.
NOTE: The data set WORK.OPTIMOUT has 6 observations and 3 variables.
NOTE: PROCEDURE OPTMODEL used (Total process time):
      real time              0.03 seconds
      cpu time               0.03 seconds
```

2.2 Assignment Problem

2.2.1 Concept of Assignment Problem

The *assignment problem* is a special class of LP problem. It deals with the situations in which resources are assigned to tasks or other work requirements. Typical examples include assignment of workers to tasks and assignment of machines to jobs. The objective is to yield an optimal matching of resources and tasks. Commonly used criteria are costs, profits, and time. The assignment problem can be described as follows. A company has a group of workers ($i = 1, 2, ..., n$) and a set of tasks ($j = 1, 2, ..., n$) to complete. The problem is how to assign n workers to n tasks at the minimum cost, c_{ij}. By introducing decision variables x_{ij} to represent the assignment of worker i to task j, the assignment model can be written as shown in Model 2.2.1.

Model 2.2.1 Standard assignment model

$$\text{Minimize } z = \sum_{i=1}^{n} \sum_{j=1}^{n} c_{ij} x_{ij} \qquad (2.2.1)$$

subject to

$$\sum_{i=1}^{n} x_{ij} = 1 \quad j = 1, 2, ..., n \qquad (2.2.2)$$

$$\sum_{j=1}^{n} x_{ij} = 1 \quad i = 1, 2, ..., n \qquad (2.2.3)$$

$$\text{All } x_{ij} \geq 0.$$

Model 2.2.1 is referred to as the *assignment model*. Objective function 2.2.1 minimizes the total cost associated with worker i performing task j.

If worker i cannot perform task j, the cost c_{ij} is considered ∞. Constraint set 2.2.2 ensures that each task is completed, and constraint set 2.2.3 ensures that every worker is assigned a task. If the total number of workers does not equal the total number of tasks, then the problem is referred to as an *unbalanced assignment problem*. A dummy worker or a dummy task is added to balance the assignment model. Because assignments via the dummy worker or dummy task are not real assignments, the costs assigned to them are 0.

In general, an assignment problem is a balanced transportation problem in which there are n origins and n destinations. Each origin has an availability of 1 unit (i.e., $s_i = 1$), and each destination has a demand of 1 unit (i.e., $d_j = 1$). Because all the right-side values of constraint sets 2.2.2 and 2.2.3 equal 1, each x_{ij} must be a nonnegative integer that is not larger than 1. As a consequence, if we solve the assignment problem as an LP model, we can guarantee that the optimal solution must be integer values and that each x_{ij} must equal 0 or 1. This proves that the assignment problem also holds the integrality property, as is the case with the transportation problem (see Section 2.1). Although the assignment model can be solved as an LP model, the resultant model can be very large. It consists of n^2 decision variables and $n!$ possible matching. A well-known method for solving the assignment model efficiently is called the *Hungarian method*.

2.2.2 Example of Assignment Problem

Figure 2.6 shows an assignment network in which there are five workers and five tasks. The cost associated with worker i performing task j are shown above the arcs, or arrows. For example, it costs 8 units of dollars for worker 1 to complete task 1. The capacity of each worker, s_i, and the demand of each task, d_j, are also shown. Because there is only one worker i available for performing a particular task j, all s_i and d_j equal 1.

This assignment network or problem can be represented by a tableau as shown in Table 2.3. The upper-right corner of each cell in the tableau represents the cost, c_{ij}.

By introducing decision variables x_{ij} to represent the assignment of worker i to task j, this assignment problem can be formulated as shown in Model 2.2.2.

Model 2.2.2 An example of formulation of assignment problem

$$\text{Minimize} \quad 8\,x_{11} + 6\,x_{12} + 2\,x_{13} + 4\,x_{14} + 3\,x_{15}$$

$$+\, 6\,x_{21} + 7\,x_{22} + 11\,x_{23} + 10\,x_{24} + 7\,x_{25}$$

$$+\, 3\,x_{31} + 5\,x_{32} + 7\,x_{33} + 6\,x_{34} + 4\,x_{35}$$

$$+\, 5\,x_{41} + 10\,x_{42} + 12\,x_{43} + 9\,x_{44} + 7\,x_{45}$$

$$+\, 7\,x_{51} + 12\,x_{52} + 5\,x_{53} + 7\,x_{54} + 8\,x_{55} \qquad (2.2.4)$$

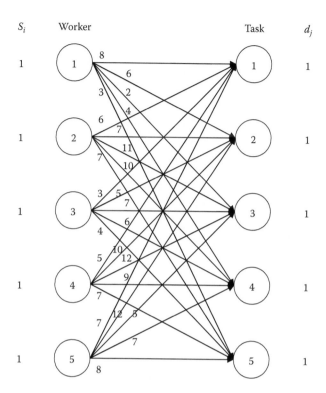

FIGURE 2.6
An assignment network.

TABLE 2.3

An Assignment Tableau

Worker i	Task j 1	2	3	4	5	s_i
1	8 x_{11}	6 x_{12}	2 x_{13}	4 x_{14}	3 x_{15}	1
2	6 x_{21}	7 x_{22}	11 x_{23}	10 x_{24}	7 x_{25}	1
3	3 x_{31}	5 x_{32}	7 x_{33}	6 x_{34}	4 x_{35}	1
4	5 x_{41}	10 x_{42}	12 x_{43}	9 x_{44}	7 x_{45}	1
5	7 x_{51}	12 x_{52}	5 x_{53}	7 x_{54}	8 x_{55}	1
d_j	1	1	1	1	1	

subject to

$$x_{11} + x_{21} + x_{31} + x_{41} + x_{51} = 1 \tag{2.2.5}$$

$$x_{12} + x_{22} + x_{32} + x_{42} + x_{52} = 1 \tag{2.2.6}$$

$$x_{13} + x_{23} + x_{33} + x_{43} + x_{53} = 1 \tag{2.2.7}$$

$$x_{14} + x_{24} + x_{34} + x_{44} + x_{54} = 1 \tag{2.2.8}$$

$$x_{15} + x_{25} + x_{35} + x_{45} + x_{55} = 1 \tag{2.2.9}$$

$$x_{11} + x_{12} + x_{13} + x_{14} + x_{15} = 1 \tag{2.2.10}$$

$$x_{21} + x_{22} + x_{23} + x_{24} + x_{25} = 1 \tag{2.2.11}$$

$$x_{31} + x_{32} + x_{33} + x_{34} + x_{35} = 1 \tag{2.2.12}$$

$$x_{41} + x_{42} + x_{43} + x_{44} + x_{45} = 1 \tag{2.2.13}$$

$$x_{51} + x_{52} + x_{53} + x_{54} + x_{55} = 1 \tag{2.2.14}$$

Constraint sets 2.2.5 to 2.2.9 ensure that each task is to be performed by exactly one worker. Constraint sets 2.2.10 to 2.2.14 ensure that each worker is to be assigned to exactly one task. Although Model 2.2.2 is an LP model, the optimal solution must be integral because the assignment model holds the integrality property.

2.2.3 ORASSIGN: SAS Code for Assignment Problem

ORASSIGN solves assignment problems, in which one set of items must be assigned to another (e.g., tasks to specific workers) at the lowest total cost (see program "sasor_2_2.sas").

Figure 2.7 illustrates the data flow in the ORASSIGN. It shows:

- The cost matrix that is required for ORASSIGN—in this case, the cost associated with worker i performing task j
- The macros (%data, %model, and %report)
- The macro variables needed to be set before running the code
- The results datasets that are available for print or can be used for further analysis

In the rest of this section, the procedure for implementing ORASSIGN, together with an example, is explained. The ORASSIGN runs three

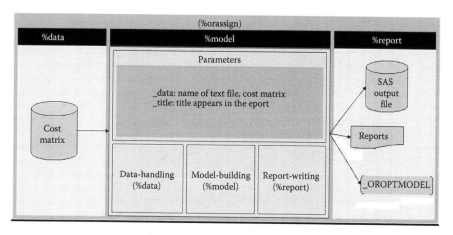

FIGURE 2.7
Data Flow in ORASSIGN.

FIGURE 2.8
An example of a dataset, cost matrix.

macros: data-handling (%data), model-building (%model), and report-writing (%report).

2.2.4 ORASSIGN: Data-Handling Macro (%data)

This part of ORASSIGN processes the data into a format that is suitable for PROC OPTMODEL. ORASSIGN requires one dataset containing names of the workers and tasks and the cost matrix. The dataset should be a .txt file, which is saved as "text tab delimited." The workers' and tasks' names must start with a letter and may contain up to 50 characters. The workers must be listed in the first column of the data file. The other columns include numeric values of tasks. These tasks can be entered into the file in any order. The first row of the data file represents the name of the tasks. An example of a data file is illustrated in Figure 2.8.

One parameter needs to set before calling the data macro:

_data: Indicates the name and location of the data file

```
* The data-handling macro;
%macro data;

 * Import text tab delimited data file to SAS data file;
 proc import
     datafile = &_data
     out = dataassign
     dbms = tab
     replace;
     getnames = yes;
 run;

%mend data;
```

2.2.5 ORASSIGN: Model-Building Macro (%model)

This part of ORASSIGN calls PROC OPTMODEL to solve the model.

The SAS marco for model-building is as following:

```
%macro model;

* Starting OPTMODEL Procedure;
proc optmodel;

* Define sets;
set < string > WORKERS;
set TASKS = {1..5};

* Define parameters;
number costprofit{WORKERS, TASKS};

* Define variables;
var X{WORKERS, TASKS} integer > = 0;

* Load the cost/profit matrix ;
read data dataassign
into WORKERS = [worker]
{t in TASKS} < costprofit[worker,t] = col("task"||t) > ;

* Define objective function;
min obj = sum{w in WORKERS, t in TASKS}
costprofit[w,t]*x[w,t];

* Define constraints;
con req_supply{w in WORKERS}:
sum{t in TASKS} x[w,t] = 1;
```

```
con req_demand{t in TASKS}:
sum{w in WORKERS} x[w,t] = 1;

* Solve the model;
solve with milp;

* Create optimum values in a SAS dataset 'optimout';
create data optimout
from [WORKERS TASKS]
 = {w in WORKERS, t in TASKS: x[w,t]^ = 0}
amount = x cp = costprofit[w,t];

* End of OPTMODEL Procedure;
quit;

%mend model;
```

It is worth noting that:

- In this code, we defined variable x as an integer variable.
- Because the assignment problem is an integer programming, we used "MILP" as a solver in the "solve" statement. This indicates that the problem should be solved using mixed-integer linear programming (MILP) as implemented in the MILP solver.
- In this example, we wanted to minimize the cost, so we used "min" statements to define the objective function. If the user wants to maximize the profit, all that is needed is simply to change the "min" statement to a "max" statement.

2.2.6 ORASSIGN: Report-Writing Macro (%report)

The ORASSIGN results include a table of workers and tasks and a total cost (or profit). This information is saved in an SAS DATA file and in print format. The SAS procedure for report-writing is as follows. One parameter needs to be set before calling this macro:

_title: Gives a title in the output of the SAS

```
%macro report;

* Report the results in a tabulated form;
proc tabulate data = optimout;
title &_title;
class WORKERS TASKS ;
var amount;
table WORKERS] = " Workers",
    TASKS*amount*sum
    / BOX = 'Assigning workers to each tasks' ;
run;

* Report the results in a tabulated form;
proc tabulate data = optimout;
title &_title;
class WORKERS TASKS ;
var cp;
table WORKERS = " Workers",
    TASKS*cp*sum
    / BOX = 'Cost-profit of workers to each tasks' ;
run;

%mend report;
```

2.2.7 ORASSIGN: Macro (%orassign)

To make the system as user friendly as possible, the %orassign macro combines the data-handling, model-building, and report-writing codes.

```
* The orassign macro for assignment problem;
%macro orassign;
 %data;
 %model;
 %report;
%mend orassign;
```

In this code, the %orassign macro is used to manage all the codes explained earlier, including data-handling, model-building, and report-writing. To get the result, the user needs to set up the parameters and run only one statement:

```
%orassign;
```

2.2.8 Instructions for Using ORASSIGN Macro

This section presents SAS code for the earlier example of assigning five workers to five tasks as shown in Table 2.1. The data are saved in file "data2_2.txt". A user needs to set the parameters as required and run the following code:

```
* The orassign macro for assignment problem;
%let _title = 'Assignment network for five workers and five
tasks.';
%let _data = 'c:\sasor\data2_2.txt';
%orassign;
```

This code produces the results based on the specified parameters and the cost matrix saved in the text file; it also produces a macro variable (_OROPTMODEL_) at termination. This variable contains a character string that indicates the status of the procedure on termination and gives the objective value at termination. Because _OROPTMODEL_ is a standard SAS macro variable, it can be used as all macro variables can be used. See the *SAS Guide to Macro Processing* for more information.

2.2.9 Sample Results from ORASSIGN Macro: Output from SAS

The code produces two tabular reports. The first table shows the assignment of tasks to workers (Figure 2.9), and the second table shows the minimum cost/maximum profit (Figure 2.10). According to the results, worker 1 is assigned to do task 3 (i.e., $x_{13} = 1$), worker 2 is assigned to do task 2 (i.e., $x_{22} = 1$), and so on. The procedure is successfully finished, and the minimum total cost is $25, as shown in Figure 2.11. The total computational time is 0.04 second.

2.2.10 Exercise

Use the codes developed in this chapter and solve the assignment problem in Table 2.4.

	TASKS				
	1	2	3	4	5
Assigning workers to each tasks	amount	amount	amount	amount	amount
	Sum	Sum	Sum	Sum	Sum
Workers					
worker1	.	.	1.00	.	.
worker2	.	1.00	.	.	.
worker3	.	.	.	.	1.00
worker4	1.00	.	.	.	.
worker5	.	.	.	1.00	.

FIGURE 2.9
Result of %orassign shows assigning tasks to workers.

	TASKS				
	1	2	3	4	5
Cost-profit of workers to each tasks	cp	cp	cp	cp	cp
	Sum	Sum	Sum	Sum	Sum
Workers					
worker1	.	.	2.00	.	.
worker2	.	7.00	.	.	.
worker3	.	.	.	.	4.00
worker4	5.00	.	.	.	.
worker5	.	.	.	7.00	.

FIGURE 2.10
Result of %orassign shows the minimum cost.

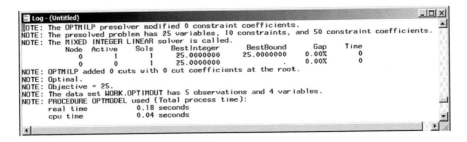

FIGURE 2.11
Log for %orassign.

TABLE 2.4

An Assignment Exercise

Worker i	Task j				s_i
	1	2	3	4	
1	6 x_{11}	2 x_{12}	4 x_{13}	3 x_{14}	1
2	6 x_{21}	7 x_{22}	10 x_{23}	8 x_{24}	1
3	5 x_{31}	7 x_{32}	6 x_{33}	4 x_{34}	1
4	7 x_{41}	5 x_{42}	7 x_{43}	8 x_{44}	1
d_j	1	1	1	1	

Solution:

- Create the data in a text file (see " data2_2_exercise.txt").
- Run the following code:

```
* SAS macro for assignment problem: solution to exercise 2.2;
%let _title = 'Assignment network: solution to exercise 2.2';
%let _data = 'c:\sasor\data2_2_exercise.txt';
%orassign;
```

The following solution is given by SAS:

Assigning workers to each tasks	TASKS			
	1	2	3	4
	amount	amount	amount	amount
	Sum	Sum	Sum	Sum
Workers				
worker1	.	1.00	.	.
worker2	1.00	.	.	.
worker3	.	.	.	1.00
worker4	.	.	1.00	.

Cost-profit of workers to each tasks	TASKS			
	1	2	3	4
	cp	cp	cp	cp
	Sum	Sum	Sum	Sum
Workers				
worker1	.	2.00	.	.
worker2	6.00	.	.	.
worker3	.	.	.	4.00
worker4	.	.	7.00	.

```
Log - (Untitled)                                                    _|□|×|
NOTE: The presolved problem has 16 variables, 8 constraints, and 32 constraint coefficients.
NOTE: The MIXED INTEGER LINEAR solver is called.
        Node  Active   Sols   BestInteger      BestBound       Gap    Time
         0      1       1     19.0000000     19.0000000      0.00%     0
         0      0       1     19.0000000          .          0.00%     0
NOTE: OPTMILP added 0 cuts with 0 cut coefficients at the root.
NOTE: Optimal.
NOTE: Objective = 19.
NOTE: The data set WORK.OPTIMOUT has 4 observations and 4 variables.
NOTE: PROCEDURE OPTMODEL used (Total process time):
        real time         0.07 seconds
        cpu time          0.04 seconds
```

2.3 Transshipment Problem

2.3.1 Concept of Transshipment Problem

The *transshipment problem* is an extension of the transportation problem. For the transportation problem, it is assumed that a commodity can only be shipped from an origin to a destination. In many real-life situations, it is also possible to distribute the commodity through the points of origins or through the points of destinations. Sometimes it might be advantageous to distribute a commodity from an origin to an intermediate, or transshipment, point before shipping it to a destination. The transshipment problem allows for these shipments.

The transshipment problem can be described as follows. A manufacturing company has a number of plants, each of which has a limited available capacity, s_i. After manufacturing, the semifinished products are delivered to the warehouses for final assembly and packaging. Finally, the finished products are shipped to the customers according to their requirements, d_j. The problem is how to fulfill each customer's order while not exceeding the capacity of any plant at the minimum cost, c_{ij}. The problem can be transformed as a conventional transportation model with $(n - b)$ origins and $(n - a)$ destinations, where n is the total number of nodes in the network (i.e., total number of plants, warehouses, and customers), a is the number of node that has supply only (or pure origin), and b is the number of nodes that has demand only (or pure destination). Any node that has both supply and demand is referred to as a *transshipment point*. The unit transportation costs, c_{ij}, are often dependent on the travel distances from node i to node j. It is assumed that the cost on a particular route of the network is directly proportional to the amount of products shipped on that route. If there is no route connecting node i and node j—or arc (i, j) does not exist—then the cost is considered ∞. The cost of delivering 1 unit of product from node i to itself is 0. By introducing decision variables x_{ij} to represent the amount of product sent from node i to node j, the transshipment model can be written as shown in Model 2.3.1.

Model 2.3.1 Standard transshipment model

$$\text{Minimize } z = \sum_{i-1}^{n-b} \sum_{j=a+1}^{n} c_{ij} x_{ij} \tag{2.3.1}$$

subject to

$$\sum_{j=a+1}^{n} x_{ij} = s_i \quad i = 1, 2, \ldots, a \tag{2.3.2}$$

$$\sum_{j=a+1}^{n} x_{ij} = s \quad i = a+1, a+2, \ldots, n-b \tag{2.3.3}$$

$$\sum_{i=1}^{n-b} x_{ij} = s \quad j = a+1, a+2, \ldots, n-b \tag{2.3.4}$$

$$\sum_{i=1}^{n-b} x_{ij} = d_j \quad j = n-b+1, n-b+2, \ldots, n \tag{2.3.5}$$

$$\text{All } x_{ij} \geq 0$$

Model 2.3.1 is referred to as the *transshipment model*. Objective function 2.3.1 minimizes the total transportation cost. Constraint set 2.3.2 is an availability constraint for the pure origin nodes ($i = 1, 2, \ldots, a$), and constraint set 2.3.3 is an availability constraint for the transshipment nodes ($i = a + 1, a + 2, \ldots, n - b$). It is assumed that all origin nodes supply the transshipment nodes. Therefore, each transshipment node will have a supply that equals the total available supply, S. Constraint set 2.3.4 is a requirement constraint for the transshipment nodes ($j = a + 1, a + 2, \ldots, n - b$), and constraint set 2.3.5 is a requirement constraint for the pure destination nodes ($j = n - b + 1, n - b + 2, \ldots, n$). For constraint set 2.3.5, if a customer also acts as a transshipment node, the customer will have a demand equal to the summation of its original demand and total available supply (i.e., $d_j + S$).

2.3.2 Example of Transshipment Problem

The following example, illustrated in Figure 2.12, shows a transshipment problem. A manufacturing company has two plants. Plant 1 can produce up to 1000 units of products per day, whereas plant 2 can produce as many as 2000 units of products per day. Instead of shipping the products directly to its customers, each product must be distributed to a warehouse first. After assembling and packaging in the warehouses, the products are shipped to the customers. The demand amounts of customer 1 and customer 2 are 1800 and 1200, respectively. The problem is how to fulfill each customer's order

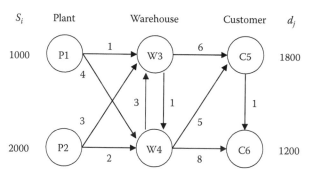

FIGURE 2.12
A transshipment network.

while not exceeding the capacity of any plant at the minimum cost, c_{ij}. Unit transportation costs, c_{ij}, are shown above the arcs (or arrows).

This transshipment model can be converted into a conventional transportation model with five origins (i.e., P1, P2, W3, W4, and C5) and four destinations (i.e., W3, W4, C5, and C6). The supplies of origins and the demands of destinations are computed as follows. For the five origins, each plant (i.e., P1 and P2) will have a supply equal to its original supply, whereas each warehouse and the first customer (i.e., W3, W4, and C5) will have a supply equal to the total available supply. For the four destinations, each warehouse (i.e., W3 and W4) will have a demand equal to the total available supply. The first customer (i.e., C5) will have a demand equal to the summation of its original demand and total available supply, whereas the second customer (i.e., C6) will have a demand equal to its original demand. The reasons for adding the total available supply to the supply and demand at each transshipment point are to ensure that the total amount of products shipped through each transshipment point will not exceed the total available supply and to also balance the transportation model.

Table 2.5 details this transshipment problem. Decision variables x_{ij} represent the amount of the products delivered from origin i to destination j. The demand of each destination, d_j, and the supply of each origin, s_i, are shown. The upper-right corner of each cell in the tableau represents the unit transportation cost, c_{ij}.

By introducing decision variables x_{ij} to represent the shipment from origin i to destination j, this transshipment problem can be formulated as shown in Model 2.3.2.

Model 2.3.2 An example of formulation of transshipment problem

$$\text{Minimize} \quad x_{13} + 4\,x_{14} + 3\,x_{23} + 2\,x_{24} + 0\,x_{33} + x_{34} + 6\,x_{35}$$

$$+ 3\,x_{43} + 0\,x_{44} + 5\,x_{45} + 8\,x_{46} + 0\,x_{55} + x_{56} \qquad (2.3.6)$$

TABLE 2.5

A Transshipment Tableau

Origin i	Destination j					s_i
	W3	**W4**	**C5**	**C6**		
P1	1 x_{13}	4 x_{14}	∞ x_{15}	∞ x_{16}		1000
P2	3 x_{23}	2 x_{24}	∞ x_{25}	∞ x_{26}		2000
W3	0 x_{33}	1 x_{34}	6 x_{35}	∞ x_{36}		3000
W4	3 x_{43}	0 x_{44}	5 x_{45}	8 x_{46}		3000
C5	∞ x_{53}	∞ x_{54}	0 x_{55}	1 x_{56}		3000
d_j	3000	3000	4800	1200		

subject to

$$x_{13} + x_{14} + x_{15} + x_{16} = 1000 \tag{2.3.7}$$

$$x_{23} + x_{24} + x_{25} + x_{26} = 2000 \tag{2.3.8}$$

$$x_{33} + x_{34} + x_{35} + x_{36} = 3000 \tag{2.3.9}$$

$$x_{43} + x_{44} + x_{45} + x_{46} = 3000 \tag{2.3.10}$$

$$x_{53} + x_{54} + x_{55} + x_{56} = 3000 \tag{2.3.11}$$

$$x_{13} + x_{23} + x_{33} + x_{43} + x_{53} = 3000 \tag{2.3.12}$$

$$x_{14} + x_{24} + x_{34} + x_{44} + x_{54} = 3000 \tag{2.3.13}$$

$$x_{15} + x_{25} + x_{35} + x_{45} + x_{55} = 4800 \tag{2.3.14}$$

$$x_{16} + x_{26} + x_{36} + x_{46} + x_{56} = 1200 \tag{2.3.15}$$

$$x_{15} = 0 \tag{2.3.16}$$

$$x_{16} = 0 \tag{2.3.17}$$

$$x_{25} = 0 \tag{2.3.18}$$

$$x_{26} = 0 \tag{2.3.19}$$

$$x_{36} = 0 \tag{2.3.20}$$

$$x_{53} = 0 \tag{2.3.21}$$

$$x_{54} = 0 \tag{2.3.22}$$

$$\text{All } x_{ij} \geq 0$$

Constraint sets 2.3.7 to 2.3.11 are the availability constraints. For example, constraint set 2.3.7 ensures that the total amount of products shipped from P1 equals its maximum production capacity. Constraint sets 2.3.12 to 2.3.15 are the requirement constraints. For example, constraint set 2.3.12 ensures that the total amount of products ordered by W3 are fulfilled. Constraint sets 2.3.16 to 2.3.22 ensure that no products are assigned to those dummy links. The optimal solution to model 2.3.2 must be integral because of the integrality property.

2.3.3 ORTRANS: SAS Code for Transshipment Problem

ORTRANS, explained in Section 2.1 (see also program "sasor_2_3.sas"), is a macro that solves transportation problems, the objective of which is to yield the minimum cost of shipment through a transportation network to destinations. The transshipment problem is an extension of the transportation problem, hence the same macro can be used for this example.

The only difference is that in the transshipment tableau (Table 2.5) some of the costs are set to be ∞. Hence in the dataset, we use a large number instead (e.g., 1E10, which is 10^{10}) because using a large number forces the user to set a variable to 0. However, such a practice introduces numerical instability, hence we added the code shown next in the program. This constraint explicity fixes the value of the corresponding variables to 0.

```
con zero{c in CUSTOMERS, s in SUPPLIERS : cost[s,c] = 1E10}:
x[s,c] = 0;
```

An example data file is seen in Figure 2.13.

Note that the following notations are used in this data file: Cstmer1 (W3), Cstmer2 (W4), Cstmer3 (C5), Cstmer4 (C6), Cstmer5 (P1), and Cstmer6 (P2).

Similar to Section 2.1, two parameters need to be set before calling the %ortrans macro:

_data: Indicates the name and location of the data file (a text tab delimited file) and contains the cost matrix

_title: Gives a title in the output of the SAS

```
* SAS procedure for transshipment problem;
%let _title = 'Example 2.3: Transshipment problem for
two suppliers, two warehouses and two customers.';
%let _data = 'c:\sasor\data2_3.txt';
%ortrans;
```

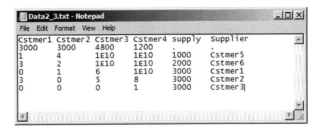

FIGURE 2.13
An example of a dataset, cost matrix for transshipment problem.

This code determines the result based on the specified parameters and the cost matrix saved in the text file and also produces a macro variable (_ORTRANS) at termination.

2.3.4 Sample Results from ORTRANS Macro: Output from SAS

The results from running this code are presented in Figure 2.14, which sorts the solutions by suppliers' name. According to the results, 1000 units of products are shipped from plant 1 (P1) to warehouse 3 (W3), 1000 units of products are shipped from warehouse 3 (W3) to customer 5 (C5), and so on. Note that the assignment of products from a transshipment point to itself (i.e., warehouse 3, warehouse 4, and customer 5) is not a real shipment. These dummy shipments simply aim at balancing the transportation model. The procedure finished successfully, and the optimal total transportation cost is $22,200, which is shown in Figure 2.15. The total computational time is less than 0.04 second.

	CUSTOMERS			
	1	2	3	4
Amount of suppliers	amount	amount	amount	amount
to customers	Sum	Sum	Sum	Sum
Suppliers				
Cstmer1	2000.00	.	1000.00	.
Cstmer2	.	1000.00	2000.00	.
Cstmer3	.	.	1800.00	1200.00
Cstmer5	1000.00	.	.	.
Cstmer6	.	2000.00	.	.

FIGURE 2.14
Result of %ortrans.

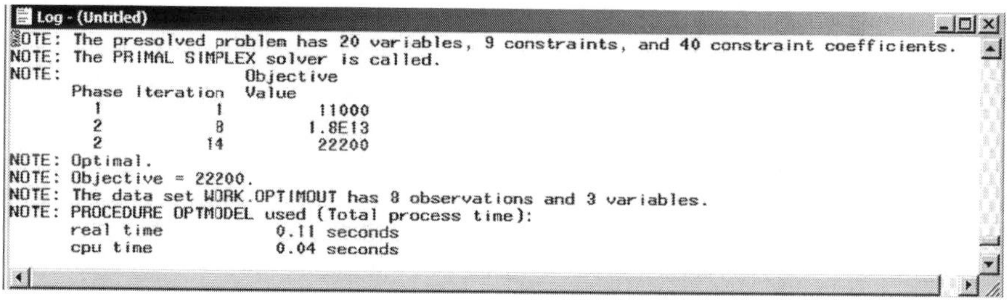

```
Log - (Untitled)                                                           _|□|x|
NOTE: The presolved problem has 20 variables, 9 constraints, and 40 constraint coefficients.
NOTE: The PRIMAL SIMPLEX solver is called.
NOTE:                     Objective
      Phase Iteration    Value
        1        1        11000
        2        8        1.8E13
        2       14        22200
NOTE: Optimal.
NOTE: Objective = 22200.
NOTE: The data set WORK.OPTIMOUT has 8 observations and 3 variables.
NOTE: PROCEDURE OPTMODEL used (Total process time):
      real time          0.11 seconds
      cpu time           0.04 seconds
```

FIGURE 2.15
Log for %ortrans.

2.3.5 Exercise

Use the codes developed in this chapter and solve the transshipment problem found in Table 2.6.

Solution:

- Create the data in a text file (see "data2_3_exercise.txt").
- Run the following code (see program "sasor_2_3_exercise.sas"):

```
* SAS macro for Transshipment problem: solution to exercise
2.3;
%let _title = 'Transshipment problem: solution to exercise
2.3.';
%let _data = 'c:\sasor\data2_3_exercise.txt';
%ortrans;
```

The following solution is given by SAS:

	CUSTOMERS				
	1	2	3	4	5
Amount of suppliers to customers	amount	amount	amount	amount	amount
	Sum	Sum	Sum	Sum	Sum
Suppliers					
Cstmer1	2000.00	1400.00	.	.	1600.00
Cstmer2	.	1600.00	1800.00	1600.00	.
Cstmer6	3000.00	.	.	.	.
Cstmer7	.	2000.00	.	.	.

TABLE 2.6

A Transshipment Exercise

	Destination j					
Origin i	W3	W4	C5	C6	C7	s_i
P1	1 x_{13}	4 x_{14}	∞ x_{15}	∞ x_{16}	∞ x_{17}	3000
P2	3 x_{23}	2 x_{24}	∞ x_{25}	∞ x_{26}	∞ x_{27}	2000
W3	0 x_{33}	1 x_{34}	5 x_{35}	7 x_{36}	4 x_{37}	5000
W4	3 x_{43}	0 x_{44}	3 x_{45}	6 x_{46}	9 x_{47}	5000
d_j	5000	5000	1800	1600	1600	

Note that the following notations are used in the data file: Cstmer1 (W3), Cstmer2 (W4), Cstmer3 (C5), Cstmer4 (C6), Cstmer5 (C7), Cstmer6 (P1), and Cstmer7 (P2).

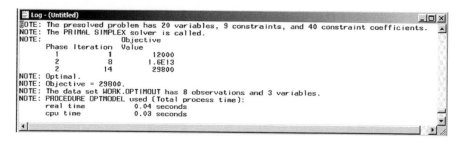

```
Log - (Untitled)                                                                    _ □ ×
NOTE: The presolved problem has 20 variables, 9 constraints, and 40 constraint coefficients.
NOTE: The PRIMAL SIMPLEX solver is called.
NOTE:                    Objective
      Phase Iteration  Value
        1        1        12000
        2        8        1.6E13
        2       14        29800
NOTE: Optimal.
NOTE: Objective = 29800.
NOTE: The data set WORK.OPTIMOUT has 8 observations and 3 variables.
NOTE: PROCEDURE OPTMODEL used (Total process time):
      real time           0.04 seconds
      cpu time            0.03 seconds
```

3

Network Models

In this chapter, we present the family of network models and demonstrate how SAS/OR® can be applied to solve the problems to optimality. The problems include minimum-cost capacitated flow, maximum flow, and shortest path problems. The problem formulations are described first. Then various SAS/OR® procedures are applied to tackle the problems with the aid of examples. Following that, result analyses are carried out. After completing this chapter, the reader will be more familiar with other SAS/OR® applications.

3.1 Minimum-Cost Capacitated Flow Problem

3.1.1 Concept of Minimum-Cost Capacitated Flow Problem

Transportation, assignment, and transshipment problems are all special cases of the *minimum-cost capacitated flow problem*. The minimum-cost capacitated flow problem aims at finding the minimum cost flow through a network while satisfying the supply and demand requirements of the origins and destinations, respectively, and also satisfying the flow restrictions through the network. However, there is no such flow restriction in transportation, assignment, and transshipment problems.

Consider a directed and connected network with n nodes in which there is at least one origin and one destination. Any node that has both supply and demand is referred to as a *transshipment point*. The minimum-cost capacitated flow problem determines how to meet the supply requirement, s_i, and the demand requirement, d_j, while not violating the capacity restrictions of any arc at the minimum cost, c_{ij}. If an arc (i, j) does not exist, the cost is considered infinite (∞). The unit cost of flow from node i to itself is 0. The lower bound on flow through arc (i, j) is L_{ij}, where $L_{ij} = 0$ if there is no lower bound. The upper bound on flow through arc (i, j) is U_{ij}, where $U_{ij} = \infty$ if there is no upper bound. By introducing decision variables x_{ij} to represent the number of units of flow sent from node i to node j through arc (i, j), the minimum-cost capacitated flow model can be written as shown in Model 3.1.1.

Model 3.1.1 Standard minimum-cost capacitated flow model

$$\text{Minimize } z = \sum_{i=1}^{n-b}\sum_{j=a+1}^{n} c_{ij}x_{ij} \tag{3.1.1}$$

subject to

$$\sum_{j=a+1}^{n} x_{ij} = s_i \quad i = 1,2,\ldots,a \tag{3.1.2}$$

$$\sum_{j=a+1}^{n} x_{ij} = S \quad i = a+1,a+2,\ldots,n-b \tag{3.1.3}$$

$$\sum_{i=1}^{n-b} x_{ij} = S \quad j = a+1,a+2,\ldots,n-b \tag{3.1.4}$$

$$\sum_{i=1}^{n-b} x_{ij} = d_j \quad j = n-b+1,n-b+2,\ldots,n \tag{3.1.5}$$

$$L_{ij} \le x_{ij} \le U_{ij} \quad i,j = 1,2,\ldots,n \tag{3.1.6}$$

$$\text{All } x_{ij} \ge 0$$

Model 3.1.1 is referred to as the *minimum-cost capacitated flow model,* which is very similar to the transshipment model presented in Section 2.3. As is the case with the transshipment problem, the minimum-cost capacitated flow problem can be transformed as a conventional transportation model with $(n-b)$ origins and $(n-a)$ destinations, where n is the total number of nodes in the network, a is the number of nodes that have supply only (or pure origin), and b is the number of nodes that have demand only (or pure destination). Objective function 3.1.1 minimizes total transportation cost. Constraint set 3.1.2 is an availability constraint for the pure origin nodes ($i = 1, 2, \ldots, a$), and constraint set 3.1.3 is an availability constraint for the transshipment nodes ($i = a + 1, a + 2, \ldots, n - b$). It is assumed that all origin nodes supply the transshipment nodes. Therefore, each transshipment node will have a supply equal to the total available supply, S. Constraint set 3.1.4 is a requirement constraint for the transshipment nodes ($j = a + 1, a + 2, \ldots, n - b$), whereas constraint set 3.1.5 is a requirement constraint for the pure destination nodes ($j = n-b + 1, n-b + 2, \ldots, n$). For constraint set 3.1.5, if a destination node also acts as a transshipment node, it will have a demand equal to the summation of its original demand and the total available supply (i.e., $d_j + S$). Constraint set 3.1.6 is the flow capacity constraint, which ensures that the flow through each arc satisfies the arc capacity restrictions.

3.1.2 Example of Minimum-Cost Capacitated Flow Problem

Figure 3.1 shows a minimum-cost capacitated flow problem. There are eight nodes; nodes 1 and node 2 are origins, and nodes 6, 7, and 8 are destinations. The amount of supply, amount of demand, unit cost on each arc, and upper bound on flow through each arc are shown in the figure.

This minimum-cost capacitated flow network (or problem) can be represented by a tableau as shown in Table 3.1. The upper-right corner of each cell in the tableau represents the unit transportation cost c_{ij}. The upper bound on flow through arc (i, j), U_{ij}, is shown in Table 3.2.

By introducing decision variables x_{ij} to represent the number of units of flow sent from node i to node j through arc (i, j), the minimum-cost capacitated flow problem can be formulated as shown in Model 3.1.2.

Model 3.1.2 Example of formulation of minimum-cost capacitated flow problem

$$\text{Minimize } 5\,x_{13} + 9999\,x_{14} + 9999\,x_{15} + 9999\,x_{16} + 9999\,x_{17} + 9999\,x_{18}$$

$$+ 9999\,x_{23} + 4\,x_{24} + 9999\,x_{25} + 9999\,x_{26} + 9999\,x_{27} + 9999\,x_{28}$$

$$+ 0\,x_{33} + 2\,x_{34} + 6\,x_{35} + 5\,x_{36} + 9999\,x_{37} + 9999\,x_{38}$$

$$+ 9999\,x_{43} + 0\,x_{44} + x_{45} + 9999\,x_{46} + 9999\,x_{47} + 2\,x_{48}$$

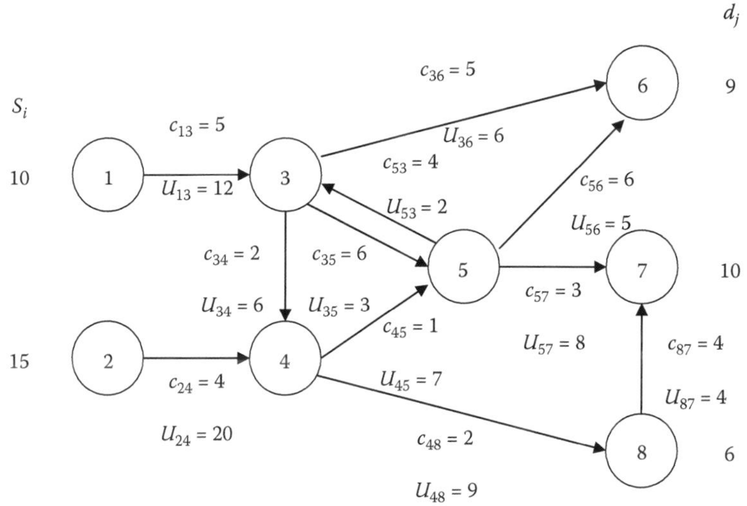

FIGURE 3.1
A minimum-cost capacitated flow network.

TABLE 3.1

A Minimum-Cost Capacitated Flow Tableau

Origin i	Destination j						s_i
	3	**4**	**5**	**6**	**7**	**8**	
1	5 x_{13}	∞ x_{14}	∞ x_{15}	∞ x_{16}	∞ x_{17}	∞ x_{18}	10
2	∞ x_{23}	4 x_{24}	∞ x_{25}	∞ x_{26}	∞ x_{27}	∞ x_{28}	15
3	0 x_{33}	2 x_{34}	6 x_{35}	5 x_{36}	∞ x_{37}	∞ x_{38}	25
4	∞ x_{43}	0 x_{44}	1 x_{45}	∞ x_{66}	∞ x_{47}	2 x_{48}	25
5	4 x_{53}	∞ x_{54}	0 x_{55}	6 x_{56}	3 x_{57}	∞ x_{58}	25
8	∞ x_{83}	∞ x_{84}	∞ x_{85}	∞ x_{86}	4 x_{87}	0 x_{88}	25
d_j	25	25	25	9	10	31	

TABLE 3.2

The Upper Bound of Each Arc

Origin i	Destination j					
	3	**4**	**5**	**6**	**7**	**8**
1	12 x_{13}	- x_{14}	- x_{15}	- x_{16}	- x_{17}	- x_{18}
2	- x_{23}	20 x_{24}	- x_{25}	- x_{26}	- x_{27}	- x_{28}
3	- x_{33}	6 x_{34}	3 x_{35}	6 x_{36}	- x_{37}	- x_{38}
4	- x_{43}	- x_{44}	7 x_{45}	- x_{46}	- x_{47}	9 x_{48}
5	2 x_{53}	- x_{54}	- x_{55}	5 x_{56}	8 x_{57}	- x_{58}
8	- x_{83}	- x_{84}	- x_{85}	- x_{86}	4 x_{87}	- x_{88}

$$+ 4\, x_{53} + 9999\, x_{54} + 0\, x_{55} + 6\, x_{56} + 3\, x_{57} + 9999\, x_{58}$$

$$+ 9999\, x_{83} + 9999\, x_{84} + 9999\, x_{85} + 9999\, x_{86} + 4\, x_{87} + 0\, x_{88} \qquad (3.1.7)$$

subject to

$$x_{13} + x_{14} + x_{15} + x_{16} + x_{17} + x_{18} = 10 \qquad (3.1.8)$$

$$x_{23} + x_{24} + x_{25} + x_{26} + x_{27} + x_{28} = 15 \qquad (3.1.9)$$

$$x_{33} + x_{34} + x_{35} + x_{36} + x_{37} + x_{38} = 25 \tag{3.1.10}$$

$$x_{43} + x_{44} + x_{45} + x_{46} + x_{47} + x_{48} = 25 \tag{3.1.11}$$

$$x_{53} + x_{54} + x_{55} + x_{56} + x_{57} + x_{58} = 25 \tag{3.1.12}$$

$$x_{83} + x_{84} + x_{85} + x_{86} + x_{87} + x_{88} = 25 \tag{3.1.13}$$

$$x_{13} + x_{23} + x_{33} + x_{43} + x_{53} + x_{83} = 25 \tag{3.1.14}$$

$$x_{14} + x_{24} + x_{34} + x_{44} + x_{54} + x_{84} = 25 \tag{3.1.15}$$

$$x_{15} + x_{25} + x_{35} + x_{45} + x_{55} + x_{85} = 25 \tag{3.1.16}$$

$$x_{16} + x_{26} + x_{36} + x_{46} + x_{56} + x_{86} = 9 \tag{3.1.17}$$

$$x_{17} + x_{27} + x_{37} + x_{47} + x_{57} + x_{87} = 10 \tag{3.1.18}$$

$$x_{18} + x_{28} + x_{38} + x_{48} + x_{58} + x_{88} = 31 \tag{3.1.19}$$

$$x_{13} \leq 12 \tag{3.1.20}$$

$$x_{24} \leq 20 \tag{3.1.21}$$

$$x_{34} \leq 6 \tag{3.1.22}$$

$$x_{35} \leq 3 \tag{3.1.23}$$

$$x_{36} \leq 6 \tag{3.1.24}$$

$$x_{45} \leq 7 \tag{3.1.25}$$

$$x_{48} \leq 9 \tag{3.1.26}$$

$$x_{53} \leq 2 \tag{3.1.27}$$

$$x_{56} \leq 5 \tag{3.1.28}$$

$$x_{57} \leq 8 \tag{3.1.29}$$

$$x_{87} \leq 4 \tag{3.1.30}$$

$$\text{All } x_{ij} \geq 0$$

Constraint sets 3.1.8 to 3.1.13 are the availability constraints. For example, constraint set 3.1.8 ensures that the total amount of flow sent from origin 1 equals its maximum amount of supply. Constraint sets 3.1.14 to 3.1.19 are the requirement constraints. For example, constraint set 3.1.17 ensures that the total amount of flow ordered by destination 6 are fulfilled. Constraint sets 3.1.20 to 3.1.30 are the flow capacity constraints. For example, constraint set 3.1.20 ensures that the flow through arc (1, 3) does not exceed its maximum arc capacity, 12. The optimal solution to Model 3.1.2 must be integral because of the integrality property.

3.1.3 ORMCFLOW: SAS Code for Minimum-Cost Capacitated Flow Problem

ORMCFLOW is a macro that solves minimum-cost capacitated flow problems, the objective of which is to find the minimum cost flow through a network while satisfying the supply and demand requirements of the origins and destinations, respectively, and also satisfying the flow restrictions through the network (see program "sasor_3_1.sas"). The primary procedure used for minimum-cost capacitated flow problem is PROC NETFLOW. A full syntax of this procedure is available in Appendix 4.

Figure 3.2 illustrates the data flow in the ORMCFLOW. It shows:

- The cost matrix that is required for ORMCFLOW, in which the cost, capacity, minimum demand, and maximum supply of any origin *i* and destination *j* are specified
- The macros (%data, %model, and %report)
- The results datasets that are available for print or can be used for further analysis

In the rest of this section, the procedure used for solving the minimum-cost capacitated flow problem (ORMCFLOW) in SAS, together with an example, is explained. The ORMCFLOW runs three macros: data-handling (%data), model-building (%model), and report-writing (%report).

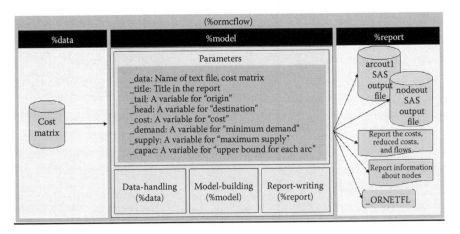

FIGURE 3.2
Data flow in ORMCFLOW.

3.1.4 ORMCFLOW: Data-Handling Macro (%data)

This part of ORMCFLOW processes the data into a format that is suitable for PROC NETFLOW. ORMCFLOW requires one dataset containing name of the origins and destinations, and the cost matrix. The dataset should be a .txt file, which is saved as "text tab delimited." The origin and destination names must start with a letter and may contain up to 50 characters. The variable names must be listed in the first row of the data file. The origin and destination names should be listed in the first and second columns, respectively; the name of these variables should be given to the macro before calling it. Note that the supply should only be given for the source nodes, whereas the demand should only be given for the sink nodes. An example of data file can be found in Figure 3.3.

One parameter needs to be set before calling the data macro:

_data: Indicates the name and location of the data file (a text tab delimited file) and contains cost matrix

```
* The data-handling macro;
%macro data;
* Import text tab delimited data file to SAS data file;
 proc import
      datafile = &_data
      out = dcost
      dbms = tab
      replace;
      getnames = yes;
 run;
%mend data;
```

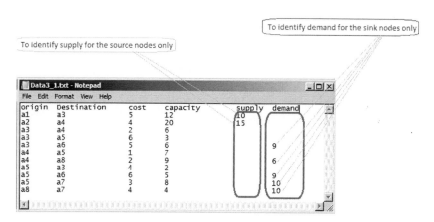

FIGURE 3.3
An example of a dataset, cost matrix.

3.1.5 ORMCFLOW: Model-Building Macro (%model)

This part of ORMCFLOW calls PROC NETFLOW to solve the model. There are six parameters prior to calling the model macro:

_tail: Defines a variable in the data file for origin

_head: Defines a variable in the data file for destination

_cost: Defines a variable in the data file for cost

_demand: Defines a variable in the data file for minimum demand

_supply: Defines a variable in the data file for maximum supply

_capac: Defines a variable in the data file for upper bound for each arc

An SAS macro for model-building is as follows:

```
* The model-building macro;
%macro model;
 proc netflow
       arcdata = dcost
       arcout = arcout1
       nodeout = nodeout1;
       tail &_tail;
       head &_head;
       cost &_cost;
       capac &_capac;
       demand &_demand;
       supply &_supply;
 run;
 %put &_ORNETFL;
%mend model;
```

3.1.6 ORMCFLOW: Report-Writing Macro (%report)

The outputs from ORMCFLOW include two reports. Report 1 contains all the information concerning arc and nonarc variables, including flows and other information regarding the current solution, the supply, and the demand, which are saved in the "arcnode" dataset. Report 2 contains all the information about nodes (supply and demand and nodal dual variable values) and other information concerning the unconstrained optimal solution, which are saved in the "nodeout" dataset. The user can define appropriate names for each of these datasets before calling %ormcflow macro:

- _arcout: Identifies the name of the SAS output file for "arcout" information
- _nodeout: Identifies the name of the SAS output file for "nodeout" information

- Another parameter needs to be set before calling this macro:
- _title: Gives a title in the output of the SAS

```
* The report-writing macro;
%macro report;
 title &_title;
 proc print
      data = arcout1;
      sum _fcost_;
 proc print
      data = nodeout1;
 run;
%mend report;
```

3.1.7 ORMCFLOW: Macro (%ormcflow)

To make the system as user friendly as possible, the %ormcflow macro combines the data-handling, model-building, and report-writing codes.

```
* A SAS macro for minimum cost capacitated flow problem;
%macro ormcflow;
 %data;
 %model;
 %report;
%mend ormcflow;
```

In this code, the %ormcflow macro is used to manage all the codes explained earlier, including data-handling, model-building, and report-writing. To get the result, the user needs to set up the parameters and run only one statement:

```
%ORMCFLOW;
```

3.1.8 Instructions for Using ORMCFLOW Macro

This section presents SAS code for the earlier example of minimum-cost capacitated flow problem with eight nodes as shown in Table 3.1. The data are saved in file "data3_1.txt."

The user needs to set the parameters as required and run the following code:

```
%let _data = 'c:/sasor/data3_1.txt';
%let _title = 'Example 3.1. Minimum cost flow problem';
%let _tail = origin;
%let _head = destination;
```

```
%let _cost = cost;
%let _capac = capacity;
%let _demand = demand;
%let _supply = supply;
%ormcflow;
```

This code determines the results based on the specified parameters and the cost matrix saved in the text file and also produces a macro variable (_ORNETFL) at termination. The user can examine the results of this macro variable, examine whether PROC NETFLOW ran correctly, and examine what error or difficulty it encountered. Because _ORNETFL is a standard SAS macro variable, it can be used as all macro variables can be used. A summary of information, including the objective value at optimum level and the status of _ORNETFL, can be seen in the log file as shown in Figure 3.4.

3.1.9 Sample Results from ORMCFLOW Macro: Output from SAS

Figure 3.5 shows all the information concerning 11 arc and nonarc variables, which among other things includes:

- The unit cost and capacity of each arc
- The supply of each tail or origin
- The demand of each head or destination
- The number of flows in each arc
- The cost of flows in each arc

Figure 3.6 shows all the information about nodes, including the supply and demand of each node, nodal dual variable values, and other information concerning the unconstrained optimal solution.

According to Figure 3.5, there are 10 units of flow from node 1 to node 3 (i.e., $x_{13} = 10$). Because the unit cost of flows through arc (1, 3) is \$5, the total cost of this particular arc is \$50. In addition, there are 15 units of flow from

```
Log - (Untitled)                                                              _|□|×|
NOTE: Number of nodes= 8 .
NOTE: Number of supply nodes= 2 .
NOTE: Number of demand nodes= 3 .
NOTE: Total supply= 25 , total demand= 25 .
NOTE: Number of arcs= 11 .
NOTE: Number of iterations performed (neglecting any constraints)= 13 .
NOTE: Of these, 2 were degenerate.
NOTE: Optimum (neglecting any constraints) found.
NOTE: Minimal total cost= 236 .
NOTE: The data set WORK.ARCOUTi has 11 observations and 13 variables.
NOTE: The data set WORK.NODEOUTi has 8 observations and 10 variables.
ERROR_STATUS=OK OPT_STATUS=OPTIMAL OBJECTIVE=236 SOLUTION=OPTIMAL
```

FIGURE 3.4
Log for %ormcflow.

VIEWTABLE: Work.Arcout1

	origin	dest	cost	capac	Arc (Nonarc) lower flow (value)	supply	demand	Arc flow or Nonarc value.	Arc flow*cost, Nonarc value*objfn coef.	Arc or Nonarc reduced cost.	Arc number.	Tail node number.	Arc or Nonarc status wrt. basi
1	a1	a3	5	12	0	10	.	10	50	.	1	1	KEY_ARC BASIC
2	a5	a3	4	2	0	.	.	0	0	10	2	5	LOWERBD NONBASIC
3	a2	a4	4	20	0	15	.	15	60	.	3	3	KEY_ARC BASIC
4	a3	a4	2	6	0	.	.	1	2	.	4	2	KEY_ARC BASIC
5	a3	a5	6	3	0	.	.	3	18	.	5	2	KEY_ARC BASIC
6	a4	a5	1	7	0	.	.	7	7	-3	6	4	UPPERBD NONBASIC
7	a3	a6	5	6	0	.	9	6	30	-7	7	2	UPPERBD NONBASIC
8	a5	a6	6	5	0	.	9	3	18	.	8	5	KEY_ARC BASIC
9	a5	a7	3	8	0	.	10	7	21	.	10	5	KEY_ARC BASIC
10	a8	a7	4	4	0	.	10	3	12	.	11	7	KEY_ARC BASIC
11	a4	a8	2	9	0	.	6	9	18	-1	9	4	UPPERBD NONBASIC

FIGURE 3.5
Result of %ormcflow in terms of "arcout."

VIEWTABLE: Work.Nodeout1

	Node name.	Supply (+) or Demand (-).	Dual variable value.	Node number.	Predecess	Traversal.	Number of successors.	Arcid, +ve (pred(i),i), -ve (i,pred(i)).	Flow (-lo) through arc arcid.	Arc length array subscript 1st arc (.,i)
1	a1	10	-99999982	1	2	4	1	-1	10	-1
2	a2	15	-99999985	3	4	8	1	-3	15	-3
3	a3	.	-99999987	2	5	1	4	-5	3	1
4	a4	.	-99999989	4	2	3	2	4	1	3
5	a5	.	-99999993	5	6	2	7	-8	3	5
6	a6	-9	-99999999	6	9	5	8	0	0	7
7	a7	-10	-99999996	8	5	7	2	10	7	10
8	a8	-6	-99999992	7	8	9	1	-11	3	9

FIGURE 3.6
Result of %ormcflow in terms of "nodeout."

node 2 to node 4 (*i.e.*, $x_{24} = 15$), and the total cost is \$60. The optimal total cost is \$236, which is shown in Figure 3.6. The total number of iterations is 8.

3.1.10 Exercise

Use the codes developed in this chapter and solve the following minimum-cost capacitated flow problem found in Tables 3.3 and 3.4.

Solution:

- Create the data in a text file (see "data3_1_exercise.txt").
- Run the following code (see program "sasor_3_1_ exercise.sas"):

```
* SAS macro for minimum cost flow problem: solution to
exercise 3.1;
%let _title = 'Minimum cost flow problem, solution to
exercise 3.1';
%let _data = 'c:/sasor/data3_1_exercise.txt';
%let _tail = origin;
```

```
%let _head = destination;
%let _cost = cost;
%let _capac = capacity;
%let _demand = demand;
%let _supply = supply;
%ormcflow;
```

The following solution is given by SAS:

	origin	Destinatio	cost	capacity	Arc (Nonarc) lower flow (value) bound.	supply	demand	Arc flow or Nonarc value.	Arc flow*cost. Nonarc value*obfn coef.	Arc or Nonarc reduced cost.	Arc number.	Tail node number.	Arc or Nonarc status wrt. basis.
1	a1	a3	1	2000	0	3000	.	2000	2000	-2	1	1	UPPERBD NONBASIC
2	a2	a3	3	1500	0	2000	.	1000	3000		2	4	KEY_ARC BASIC
3	a4	a3	3	1000	0		.	0	0	4	3	3	LOWERBD NONBASIC
4	a1	a4	4	1500	0	3000	.	1000	4000		4	1	KEY_ARC BASIC
5	a2	a4	2	1000	0	2000	.	1000	2000	-2	5	4	UPPERBD NONBASIC
6	a3	a4	1	1000	0		.	1000	1000	0	6	2	UPPERBD NONBASIC
7	a3	a5	5	1600	0		1900	200	1000		7	2	KEY_ARC BASIC
8	a4	a5	3	1600	0		1800	1600	4800	-1	8	3	UPPERBD NONBASIC
9	a3	a6	7	1600	0		1600	200	1400		9	2	KEY_ARC BASIC
10	a4	a6	6	1600	0		1500	1400	8400		10	3	KEY_ARC BASIC
11	a3	a7	4	1600	0		1600	1600	6400	-3	11	2	UPPERBD NONBASIC
12	a4	a7	9	1600	0		1600	0	0	3	12	3	LOWERBD NONBASIC

VIEWTABLE: Work.Arcout1

TABLE 3.3

A Minimum-Cost Capacitated Flow Exercise

			Destination j			
Origin i	3	4	5	6	7	s_i
1	1 x_{13}	4 x_{14}	∞ x_{15}	∞ x_{16}	∞ x_{17}	3000
2	3 x_{23}	2 x_{24}	∞ x_{25}	∞ x_{26}	∞ x_{27}	2000
3	0 x_{33}	1 x_{34}	5 x_{35}	7 x_{36}	4 x_{37}	5000
4	3 x_{43}	0 x_{44}	3 x_{45}	6 x_{46}	9 x_{47}	5000
dj	5000	5000	1800	1600	1600	

TABLE 3.4

The Upper Bound of Each Arc

		Destination j			
Origin i	3	4	5	6	7
1	2000 x_{13}	1500 x_{14}	- x_{15}	- x_{16}	- x_{17}
2	1500 x_{23}	1000 x_{24}	- x_{25}	- x_{26}	- x_{27}
3	- x_{33}	1000 x_{34}	1600 x_{35}	1600 x_{36}	1600 x_{37}
4	1000 x_{43}	- x_{44}	1600 x_{45}	1600 x_{46}	1600 x_{47}

3.2 Maximum Flow Problem

3.2.1 Concept of Maximum Flow Problem

The *maximum flow problem* is a special case of the minimum-cost capacitated flow problem. There are three major differences between them. First, the cost per unit of flow through the arc (i, j) or c_{ij} is not considered in the maximum flow problem. Second, there must be exactly one starting point (i.e., the source) and exactly one terminal point (i.e., the sink) in the maximum flow problem. Third, the objective of the maximum flow problem is to yield the maximum amount of flow from a source to a sink.

Consider a directed and connected network with n nodes plus a source node, s, and a sink node, t. All the remaining n nodes are transshipment points. The objective of the maximum flow problem is to maximize the total amount of flow entering the network, x_{sj}, while not violating the capacity restrictions of any arc. The lower bound on flow through arc (i, j) is L_{ij}, where $L_{ij} = 0$ if there is no lower bound. The upper bound on flow through arc (i, j) is U_{ij}, where $U_{ij} = \infty$ if there is no upper bound. By introducing variables x_{ij} to represent the number of units of flow sent from node i to node j through arc (i, j), the maximum flow model can be written as shown in Model 3.2.1.

Model 3.2.1 Standard maximum flow model

$$\text{Maximize } z = \sum_{j=1}^{n} x_{ij} \qquad (3.2.1)$$

subject to

$$\sum_{j} x_{ij} - \sum_{j} x_{ij} = b_i \quad i = 1, 2, \ldots, n \qquad (3.2.2)$$

$$L_{ij} \leq x_{ij} \leq U_{ij} \quad i, j = 1, 2, \ldots, n \qquad (3.2.3)$$

$$\text{All } x_{ij} \geq 0$$

Model 3.2.1 is referred to as the *standard maximum flow model*. Objective function 3.2.1 maximizes the total amount of flow entering the directed and connected network. All arcs are directional, or one way, which means that flow through an arc is allowed only in the direction shown by the arrowhead. If bidirectional flows are permitted on a particular arc, this would be represented by a pair of arcs pointing in opposite directions. Constraint set 3.2.2 is the flow balance constraint for all transshipment points, in which the first summation represents the total flow into node i while the second summation represents the total flow out of node i, where $j = s, 1, 2, \ldots, n, t$. So this constraint set ensures that the difference of two summations, or the net flow at node i, must equal b_i. Because all n nodes are transshipment points, b_i has a zero value. Constraint set 3.2.3, the flow capacity constraint, ensures that the flow through each arc satisfies the arc capacity restrictions.

3.2.2 Example of Maximum Flow Problem

Figure 3.7 shows a maximum flow problem, which is an extension of the example shown in Figure 3.1. Because there is exactly one source and one sink in the maximum flow problem, nodes 1, 2, 6, 7, and 8 become transshipment points. Node 0 is called the *source node* because there are flows out of it but no flow into it. Node 9 is called the *sink node* because there are flows into it but no flow out of it. The upper and lower bounds on flow through each arc are shown in Figure 3.7 and Tables 3.5 and 3.6.

By introducing decision variables x_{ij} to represent the number of units of flow sent from node i to node j through arc (i, j), the maximum flow problem can be formulated as shown in Model 3.2.2.

Model 3.2.2 Example of formulation of maximum flow problem

$$\text{Maximize} \quad x_{01} + x_{02} \tag{3.2.4}$$

subject to

$$x_{01} - x_{13} = 0 \tag{3.2.5}$$

$$x_{02} - x_{24} = 0 \tag{3.2.6}$$

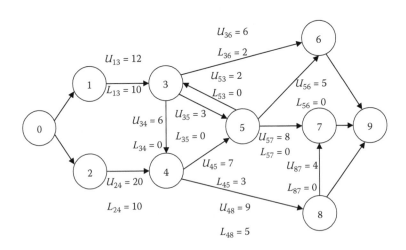

FIGURE 3.7
A maximum flow network.

TABLE 3.5

The Upper Bound of Each Arc

Origin i	Destination j					
	3	4	5	6	7	8
1	12 x_{13}	- x_{14}	- x_{15}	- x_{16}	- x_{17}	- x_{18}
2	- x_{23}	20 x_{24}	- x_{25}	- x_{26}	- x_{27}	- x_{28}
3	- x_{33}	6 x_{34}	3 x_{35}	6 x_{36}	- x_{37}	- x_{38}
4	- x_{43}	- x_{44}	7 x_{45}	- x_{46}	- x_{47}	9 x_{48}
5	2 x_{53}	- x_{54}	- x_{55}	5 x_{56}	8 x_{57}	- x_{58}
8	- x_{83}	- x_{84}	- x_{85}	- x_{86}	4 x_{87}	- x_{88}

TABLE 3.6

The Lower Bound of Each Arc

Origin i	Destination j					
	3	4	5	6	7	8
1	10 x_{13}	- x_{14}	- x_{15}	- x_{16}	- x_{17}	- x_{18}
2	- x_{23}	10 x_{24}	- x_{25}	- x_{26}	- x_{27}	- x_{28}
3	- x_{33}	0 x_{34}	0 x_{35}	2 x_{36}	- x_{37}	- x_{38}
4	- x_{43}	- x_{44}	3 x_{45}	- x_{46}	- x_{47}	5 x_{48}
5	0 x_{53}	- x_{54}	- x_{55}	0 x_{56}	0 x_{57}	- x_{58}
8	- x_{83}	- x_{84}	- x_{85}	- x_{86}	0 x_{87}	- x_{88}

$$x_{13} - x_{34} - x_{35} - x_{36} + x_{53} = 0 \qquad (3.2.7)$$

$$x_{24} + x_{34} - x_{45} - x_{48} = 0 \qquad (3.2.8)$$

$$x_{35} + x_{45} - x_{53} - x_{56} - x_{57} = 0 \qquad (3.2.9)$$

$$x_{36} + x_{56} - x_{69} = 0 \qquad (3.2.10)$$

$$x_{57} + x_{87} - x_{79} = 0 \qquad (3.2.11)$$

$$x_{48} - x_{87} - x_{89} = 0 \qquad (3.2.12)$$

$$x_{13} \leq 12 \qquad (3.2.13)$$

$$x_{24} \leq 20 \qquad (3.2.14)$$

$$x_{34} \leq 6 \qquad (3.2.15)$$

$$x_{35} \leq 3 \qquad (3.2.16)$$

$$x_{36} \leq 6 \qquad (3.2.17)$$

$$x_{45} \leq 7 \qquad (3.2.18)$$

$$x_{48} \leq 9 \qquad (3.2.19)$$

$$x_{53} \leq 2 \qquad (3.2.20)$$

$$x_{56} \leq 5 \qquad (3.2.21)$$

$$x_{57} \leq 8 \qquad (3.2.22)$$

$$x_{87} \leq 4 \qquad (3.2.23)$$

$$x_{13} \geq 10 \qquad (3.2.24)$$

$$x_{24} \geq 10 \qquad (3.2.25)$$

$$x_{34} \geq 0 \qquad (3.2.26)$$

$$x_{35} \geq 0 \qquad (3.2.27)$$

$$x_{36} \geq 2 \qquad (3.2.28)$$

$$x_{45} \geq 3 \qquad (3.2.29)$$

$$x_{48} \geq 5 \qquad (3.2.30)$$

$$x_{53} \geq 0 \qquad (3.2.31)$$

$$x_{56} \geq 0 \qquad (3.2.32)$$

$$x_{57} \geq 0 \qquad (3.2.33)$$

$$x_{87} \geq 0 \qquad (3.2.34)$$

$$\text{All } x_{ij} \geq 0$$

Constraint sets 3.2.5 to 3.2.12 are the flow balance constraints. For example, constraint set 3.2.5 ensures that the total flow into node 1 (i.e., x_{01}) is the same as the total flow out of node 1 (i.e., x_{13}). Herein, "+" is added to the flows into a node, whereas "−" is added to the flows out of a node. Constraint sets 3.2.13 to 3.2.23 are the maximum flow capacity constraints. For example, constraint set 3.2.13 ensures that the flow through arc (1, 3) must not exceed its maximum arc capacity, 12. Constraint sets 3.2.24 to 3.2.34 are the minimum flow requirement constraints. For example, constraint set 3.2.2 ensures that the flow through arc (1, 3) must at least meet its minimum arc requirement, 10. In cases where every U_{ij} and L_{ij} have integer values, the optimal solution to Model 3.2.2 must be integral.

3.2.3 ORMAXFLOW: SAS Code for Maximum Flow Problem

ORMCFLOW, explained in Section 3.1, is a macro that can also be used to solve maximum flow problems, which is a special case of the minimum-cost capacitated flow problem. Hence, we use a similar macro with some minor changes to solve the maximum flow problem. The new macro is called ORMAXFLOW (see program "sasor_3_2.sas").

Figure 3.8 illustrates the data flow in the ORMAXFLOW. It shows:

- The cost matrix that is required for ORMAXFLOW, in which the cost, capacity, minimum demand, and maximum supply of any origin i and destination j are specified
- The macros (%data, %model, and %report)
- The results datasets that are available for print or can be used for further analysis

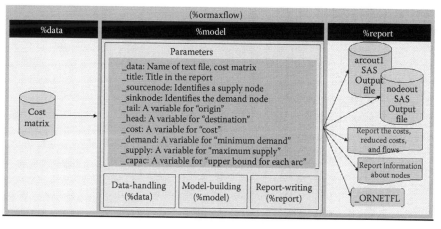

FIGURE 3.8
Data flow in ORMAXFLOW.

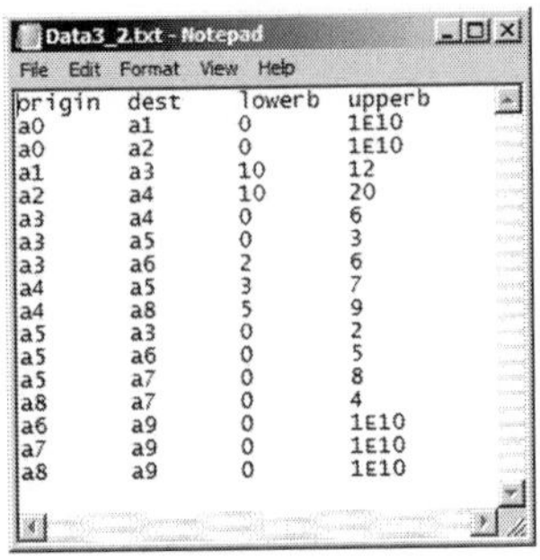

Data3_2.txt - Notepad

origin	dest	lowerb	upperb
a0	a1	0	1E10
a0	a2	0	1E10
a1	a3	10	12
a2	a4	10	20
a3	a4	0	6
a3	a5	0	3
a3	a6	2	6
a4	a5	3	7
a4	a8	5	9
a5	a3	0	2
a5	a6	0	5
a5	a7	0	8
a8	a7	0	4
a6	a9	0	1E10
a7	a9	0	1E10
a8	a9	0	1E10

FIGURE 3.9
An example of a dataset with lower and upper bounds for maximum flow problem.

In the rest of this section, the procedure for solving a maximum flow problem (ORMAXFLOW) in SAS, together with an example, is explained. The ORMAXFLOW runs three macros: data-handling (%data), model-building (%model), and report-writing (%report).

The only difference between the problem in this section and the one found in Section 3.1 is that the dataset contains both lower flow bound and upper flow bound. An example of a dataset is shown in Figure 3.9.

Similar to Section 3.1, some parameters need to be set before calling the %ormaxflow macro (see program "sasor_3_2.sas").:

```
* SAS procedure for maximum flow problem;
%let _data = 'c:/sasor/data3_2.txt';
%let _title = 'Example 3.2. Maximum flow problem ';
%let _sourcenode = 'a0';
%let _sinknode = 'a9';
%let _tail = origin;
%let _head = dest;
%let _lowerb = lowerb;
%let _capac = upperb;
% ormaxflow;
```

This code determines the results based on the specified parameters and the lower and upper bounds saved in the text file. Note that there are two further macro variables (_sourcenode and _sinknode) to identify the source node and the sink node. This code also produces a macro variable (_ORNETFL) at termination.

The complete SAS code for maximum flow problem contains three macros (%data, %model, and %report) as listed next. In this code, we also used the option "maxflow" in the PROC NETFLOW; this option indicates that the procedure should maximize the flow in the program.

```
* The data-handling macro;
%macro data;
* Import text tab delimited data file to SAS data file;
 proc import
      datafile = &_data
      out = dcost
      dbms = tab
      replace;
      getnames = yes;
 run;
%mend data;
```

```
* The model-building macro;
%macro model;
 proc netflow
      maxflow
      arcdata = dcost
      source = &_sourcenode
      sink = &_sinknode
      arcout = arcout1
      nodeout = nodeout1;
      minflow &_lowerb;
      capac &_capac;
      tail &_tail;
      head &_head;
 run;
 %put &_ORNETFL;
%mend model;
```

```
* The report-writing macro;
%macro report;
 title &_title ', how to get from ' &_sourcenode ' to '
&_sinknode ;
 proc print
      data = arcout1;
      sum _fcost_;
 proc print
      data = nodeout1;
 run;
%mend report;
```

```
* A SAS macro for maximum flow problem;
%macro ormaxflow;
 %data;
 %model;
 %report;
%mend ormaxflow;
```

```
* SAS procedure for maximum flow problem;
%let _data = 'c:/sasor/data3_2.txt';
%let _title = 'Example 3.2. Maximum flow problem ';
%let _sourcenode = 'a1';
%let _sinknode = 'a8';
%let _tail = origin;
%let _head = dest;
%let _lowerb = lowerb;
%let _capac = upperb;
%ormaxflow;
```

3.2.4 Sample Results from ORMAXFLOW Macro: Output from SAS

Figure 3.10 shows all the information concerning 16 arc and nonarc variables, including;

- The unit cost of each arc
- The upper and lower bounds of each arc
- The supply of the tail or origin
- The demand of the head or destination
- The number of flows in each arc
- The cost of flows in each arc.

Figure 3.11 shows all the information about nodes, including the supply and demand of each node, the nodal dual variable values, and other information concerning the unconstrained optimal solution.

According to Figure 3.10, there are 12 units of flow from node 0 to node 1 (i.e., $x_{01} = 12$). Because the unit cost of flows is not considered in the maximum

	dest	Arc cost or Nonarc objfn coef.	upperb	lowerb	Supply of tail node.	Demand of head node.	Arc flow or Nonarc value.	Arc flow*cost, Nonarc value*objfn coef	Arc or Nonarc reduced cost.	Arc number.	Tail node number.	Arc or Nonarc status wrt. base.
1	a1	0	10000000000	0	99999998	.	12	0	.	1	1	KEY_ARC EASIC
2	a2	0	10000000000	0	99999998	.	13	0	.	2	1	KEY_ARC EASIC
3	a3	0	12	10	.	.	12	0	0	3	2	UPPERBD NONBASIC
4	a3	0	2	0	.	.	0	0	1	4	6	LOWERBD NONBASIC
5	a4	0	20	10	.	.	13	0	.	5	3	KEY_ARC EASIC
6	a4	0	6	0	.	.	3	0	.	6	4	KEY_ARC EASIC
7	a5	0	3	0	.	.	3	0	-1	7	4	UPPERBD NONBASIC
8	a5	0	7	3	.	.	7	0	-1	8	5	UPPERBD NONBASIC
9	a6	0	6	2	.	.	6	0	-1	9	4	UPPERBD NONBASIC
10	a6	0	5	0	.	.	2	0	.	10	6	KEY_ARC EASIC
11	a7	0	8	0	.	.	8	0	0	12	6	UPPERBD NONBASIC
12	a7	0	4	0	.	.	4	0	0	13	8	UPPERBD NONBASIC
13	a8	0	9	5	.	.	9	0	-1	11	5	UPPERBD NONBASIC
14	a9	0	10000000000	0	99999998	.	8	0	.	14	7	KEY_ARC EASIC
15	a9	0	10000000000	0	99999998	.	12	0	.	15	9	KEY_ARC EASIC
16	a9	0	10000000000	0	99999998	.	5	0	.	16	8	KEY_ARC EASIC

FIGURE 3.10
Result of %ormaxflow in terms of "arcout."

flow problem, the total cost of any arc is 0. In addition, there are 13 units of flow from node 0 to node 2 (i.e., $x_{02} = 13$). The optimal total amount of flows is 25, which is shown in Figure 3.12. The total number of iterations is 16.

3.2.5 Exercise

Use the codes developed in this chapter and solve the following maximum flow problem found in Tables 3.7 and 3.8.

	Node name.	Supply (+) or Demand (-).	Dual variable value.	Node number.	Predecessor.	Traversal.	Number of successors.	Arcid. +ve (pred(i).i). -ve (i.pred()).	Flow (-lo) through arc arcid.	Arc length array subscript 1st arc (.i)
1	a0	99999998	-99999998	1	10	3	5	-17	99999973	-1
2	a1	.	-99999998	2	1	8	1	1	12	1
3	a2	.	-99999998	3	1	5	3	2	13	2
4	a3	.	-99999998	4	5	2	1	-6	3	3
5	a4	.	-99999998	5	3	4	2	5	3	5
6	a5	.	-99999999	6	7	11	1	-10	2	7
7	a6	.	-99999999	7	10	6	2	-14	8	9
8	a7	.	-99999999	9	11	10	10	0	0	12
9	a8	.	-99999999	8	10	7	1	-16	5	11
10	a9	-99999998	-99999999	10	9	1	9	15	12	14

FIGURE 3.11
Result of %ormaxflow in terms of "nodeout."

```
Log - (Untitled)
NOTE: Number of nodes= 10 .
NOTE: Number of arcs= 17 .
NOTE: Number of iterations performed (neglecting any constraints)= 16 .
NOTE: Of these, 1 were degenerate.
NOTE: Maximal flow= 25 .
NOTE: Optimum (neglecting any constraints) found.
NOTE: Minimal total cost= 0 .
NOTE: The data set WORK.ARCOUT1 has 16 observations and 13 variables.
NOTE: The data set WORK.NODEOUT1 has 10 observations and 10 variables.
ERROR_STATUS=OK OPT_STATUS=OPTIMAL OBJECTIVE=0 MAXFLOW=25 SOLUTION=OPTIMAL

NOTE: PROCEDURE NETFLOW used (Total process time):
      real time           0.53 seconds
      cpu time            0.10 seconds
```

FIGURE 3.12
Log for %ormaxflow.

TABLE 3.7

The Upper Bound of Each Arc

Origin i	Destination j				
	3	4	5	6	7
1	2000 x_{13}	1500 x_{14}	- x_{15}	- x_{16}	- x_{17}
2	1500 x_{23}	1000 x_{24}	- x_{25}	- x_{26}	- x_{27}
3	- x_{33}	1000 x_{34}	1600 x_{35}	1600 x_{36}	1600 x_{37}
4	1000 x_{43}	- x_{44}	1600 x_{45}	1600 x_{46}	1600 x_{47}

TABLE 3.8

The Lower Bound of Each Arc

Origin i	\multicolumn{5}{c}{Destination j}				
	3	**4**	**5**	**6**	**7**
1	0	0	–	–	–
	x_{13}	x_{14}	x_{15}	x_{16}	x_{17}
2	0	0	–	–	–
	x_{23}	x_{24}	x_{25}	x_{26}	x_{27}
3	–	0	0	0	0
	x_{33}	x_{34}	x_{35}	x_{36}	x_{37}
4	0	–	0	0	0
	x_{43}	x_{44}	x_{45}	x_{46}	x_{47}

Solution:

- Create the data in a text file (see "data3_2_exercise.txt").
- Run the following code (see program "sasor_3_2_ exercise.sas"):

```
* A SAS procedure for maximum flow problem: solution to
exercise 3.2.;
%let _title = 'Maximum flow problem: solution to exercise 3.2';
option nodate;
%let _data = 'c:/sasor/data3_2_exercise.txt';
%let _sourcenode = 'a0';
%let _sinknode = 'a8';
%let _tail = origin;
%let _head = dest;
%let _lowerb = lowerb;
%let _capac = upperb;
%ormaxflow;
```

The following solution is given by SAS:

	origin	dest	Arc cost or Nonarc obfn coef.	upperb	lowerb	Supply of tail node.	Demand of head node.	Arc flow or Nonarc value.	Arc flow-cost, Nonarc value-obfn cost	Arc or Nonarc reduced cost	Arc number.	Tail node number.	Arc or Nonarc status wrt. basis.
1	a0	a1	0	10000000000	0	99999998	.	3500	0	.	1	1	KEY_ARC BASIC
2	a0	a2	0	10000000000	0	99999998	.	2500	0	.	2	1	KEY_ARC BASIC
3	a1	a3	0	2000	0	.	.	2000	0	-1	3	2	UPPERBD NONBASIC
4	a2	a3	0	1500	0	.	.	1500	0	-1	4	3	UPPERBD NONBASIC
5	a4	a3	0	1000	0	.	.	0	0	0	5	5	LOWERBD NONBASIC
6	a1	a4	0	1500	0	.	.	1500	0	-1	6	2	UPPERBD NONBASIC
7	a2	a4	0	1000	0	.	.	1000	0	-1	7	3	UPPERBD NONBASIC
8	a3	a4	0	1000	0	.	.	700	0	.	8	4	KEY_ARC BASIC
9	a3	a5	0	1000	0	.	.	1000	0	0	9	4	UPPERBD NONBASIC
10	a4	a5	0	1600	0	.	.	1600	0	0	10	5	UPPERBD NONBASIC
11	a3	a6	0	1600	0	.	.	1600	0	0	11	4	UPPERBD NONBASIC
12	a4	a6	0	1600	0	.	.	1600	0	0	12	5	UPPERBD NONBASIC
13	a3	a7	0	1600	0	.	.	200	0	.	13	4	KEY_ARC BASIC
14	a4	a7	0	1600	0	.	.	0	0	0	14	5	LOWERBD NONBASIC
15	a5	a8	0	10000000000	0	.	1E8	2600	0	.	15	6	KEY_ARC BASIC
16	a6	a8	0	10000000000	0	.	1E8	3200	0	.	16	7	KEY_ARC BASIC
17	a7	a8	0	10000000000	0	.	1E8	200	0	.	17	8	KEY_ARC BASIC

3.3 Shortest Path Problem

3.3.1 Concept of Shortest Path Problem

The *shortest path problem* aims at finding a shortest path between a starting node and a terminal node through a network. The problem can be regarded as a special case of the transshipment problem, which was discussed in Section 2.3. Consider a directed and connected network with n nodes in which there is exactly one origin and one destination. All the remaining nodes are transshipment points. The shortest path problem is to minimize the cost of shipping 1 unit of product from node i to node j, c_{ij}. If arc (i, j) exists, the unit transportation cost, c_{ij}, is the same as the length of such an arc. Otherwise, $c_{ij} = \infty$. The cost of delivering 1 unit of product from node i to itself is 0. As mentioned in Section 2.3, the number of nodes that has supply only, or pure origin, is denoted as a, whereas the number of node, that has demand only, or pure destination, is denoted as b. Because there is exactly one origin and one destination in the shortest path problem, both a and b equal 1. By introducing decision variables x_{ij} to represent the flow from node i to node j, the shortest path model can be written as shown in Model 3.3.1.

Model 3.3.1 Standard shortest path model

$$\text{Minimize } z = \sum_{i=j}^{n-b} \sum_{j=a+1}^{n} c_{ij} x_{ij} \tag{3.3.1}$$

subject to

$$\sum_{j=a+1}^{n} x_{ij} = 1 \quad i = 1, 2, \ldots, n-b \tag{3.3.2}$$

$$\sum_{i=1}^{n-b} x_{ij} = 1 \quad j = a+1, a+2, \ldots, n \tag{3.3.3}$$

$$\text{All } x_{ij} \geq 0.$$

Model 3.3.1 is referred to as the *shortest path model*. Objective function 3.3.1 finds a path that connects the origin and the destination and requires the minimum total transportation cost. Constraint set 3.3.2 is an availability constraint, which guarantees that the total maximum amount of products shipped from node i equals 1. Constraint set 3.3.3 is a requirement constraint, which ensures that the total maximum amount of products received by node j equals 1.

3.3.2 Example of Shortest Path Problem

Figure 3.13 shows a shortest path network with seven nodes. Node 1 is an origin, whereas node 7 is a destination. All the remaining nodes are transshipment points. The unit transportation cost is shown above each arc.

The shortest path network is a special case of transshipment problem and can be transformed into a tableau as shown in Table 3.9. Nodes 1 to 6 can be regarded as origins, whereas nodes 2 to 7 can be treated as

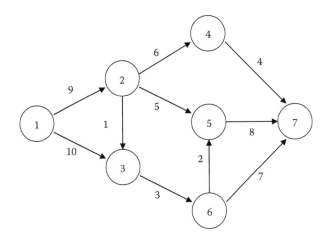

FIGURE 3.13
A shortest path network.

TABLE 3.9

A Shortest Path Tableau

Origin i	Destination j						s_i
	2	3	4	5	6	7	
1	9 $\quad x_{12}$	10 $\quad x_{13}$	∞ $\quad x_{14}$	∞ $\quad x_{15}$	∞ $\quad x_{16}$	∞ $\quad x_{17}$	1
2	0 $\quad x_{22}$	1 $\quad x_{23}$	6 $\quad x_{24}$	5 $\quad x_{25}$	∞ $\quad x_{26}$	∞ $\quad x_{27}$	1
3	∞ $\quad x_{32}$	0 $\quad x_{33}$	∞ $\quad x_{34}$	∞ $\quad x_{35}$	3 $\quad x_{36}$	∞ $\quad x_{37}$	1
4	∞ $\quad x_{42}$	∞ $\quad x_{43}$	0 $\quad x_{44}$	∞ $\quad x_{45}$	∞ $\quad x_{46}$	4 $\quad x_{47}$	1
5	∞ $\quad x_{52}$	∞ $\quad x_{53}$	∞ $\quad x_{54}$	0 $\quad x_{55}$	∞ $\quad x_{56}$	8 $\quad x_{57}$	1
6	∞ $\quad x_{62}$	∞ $\quad x_{63}$	∞ $\quad x_{64}$	2 $\quad x_{65}$	0 $\quad x_{66}$	7 $\quad x_{67}$	1
d_j	1	1	1	1	1	1	

destinations. Decision variables x_{ij} represent the quantity of the products delivered from origin i to destination j. The demand of each destination is denoted as d_j, whereas the supply of each origin is denoted as s_i. Because we want to ship 1 unit of product from node 1 to node 7, all d_j and s_i equal 1. The upper-right corner of each cell in the tableau represents the unit transportation cost, c_{ij}. If arc (i, j) does not exist, the cost c_{ij} is ∞. For any arc (i, i), the cost c_{ii} is 0.

By introducing decision variables x_{ij} to represent the shipment from origin i to destination j, this shortest path problem can be formulated as shown in Model 3.3.2.

Model 3.3.2 Example of formulation of shortest path problem

$$\text{Minimize} \quad 9\,x_{12} + 10\,x_{13} + 9999\,x_{14} + 9999\,x_{15} + 9999\,x_{16} + 9999\,x_{17}$$

$$+\,0\,x_{22} + x_{23} + 6\,x_{24} + 5\,x_{25} + 9999\,x_{26} + 9999\,x_{27}$$

$$+\,9999\,x_{32} + 0\,x_{33} + 9999\,x_{34} + 9999\,x_{35} + 3\,x_{36} + 9999\,x_{37}$$

$$+\,9999\,x_{42} + 9999\,x_{43} + 0\,x_{44} + 9999\,x_{45} + 9999\,x_{46} + 4\,x_{47}$$

$$+\,9999\,x_{52} + 9999\,x_{53} + 9999\,x_{54} + 0\,x_{55} + 9999\,x_{56} + 8\,x_{57}$$

$$+\,9999\,x_{62} + 9999\,x_{63} + 9999\,x_{64} + 2\,x_{65} + 0\,x_{66} + 7\,x_{67} \quad (3.3.4)$$

subject to

$$x_{12} + x_{13} + x_{14} + x_{15} + x_{16} + x_{17} = 1 \qquad (3.3.5)$$

$$x_{22} + x_{23} + x_{24} + x_{25} + x_{26} + x_{27} = 1 \qquad (3.3.6)$$

$$x_{32} + x_{33} + x_{34} + x_{35} + x_{36} + x_{37} = 1 \qquad (3.3.7)$$

$$x_{42} + x_{43} + x_{44} + x_{45} + x_{46} + x_{47} = 1 \qquad (3.3.8)$$

$$x_{52} + x_{53} + x_{54} + x_{55} + x_{56} + x_{57} = 1 \qquad (3.3.9)$$

$$x_{62} + x_{63} + x_{64} + x_{65} + x_{66} + x_{67} = 1 \qquad (3.3.10)$$

$$x_{12} + x_{22} + x_{32} + x_{42} + x_{52} + x_{62} = 1 \qquad (3.3.11)$$

$$x_{13} + x_{23} + x_{33} + x_{43} + x_{53} + x_{63} = 1 \qquad (3.3.12)$$

$$x_{14} + x_{24} + x_{34} + x_{44} + x_{54} + x_{64} = 1 \tag{3.3.13}$$

$$x_{15} + x_{25} + x_{35} + x_{45} + x_{55} + x_{65} = 1 \tag{3.3.14}$$

$$x_{16} + x_{26} + x_{36} + x_{46} + x_{56} + x_{66} = 1 \tag{3.3.15}$$

$$x_{17} + x_{27} + x_{37} + x_{47} + x_{57} + x_{67} = 1 \tag{3.3.16}$$

$$\text{All } x_{ij} \geq 0$$

Constraint sets 3.3.5 to 3.3.10 are the availability constraints. For example, constraint set 3.3.5 ensures that the total maximum amount of products shipped from node 1 equals 1. Constraint sets 3.3.11 to 3.3.16 are the requirement constraints. For example, constraint set 3.3.11 ensures that the total maximum amount of products received by node 2 equals 1. Because of the integrality property that the transshipment problem has, we can be sure that this shortest path flow through each arc will be 0 or 1, even when we solve the problem as the LP problem.

3.3.3 ORSHORTPATH: SAS Code for Shortest Path Problem

ORMCFLOW (see Section 3.1) is a macro that can also be used to solve the shortest path problem, which is a special case of the minimum-cost capacitated flow problem, and it aims at finding a shortest path between a starting node and a terminal node. Hence, we use a similar macro with some minor changes to solve the shortest path problem. The new macro is called ORSHORTPATH (see program "sasor_3_3.sas").

The only difference is that the dataset only contains the name of origins and destinations and the cost of each arc. An example of such a dataset is shown in Figure 3.14.

FIGURE 3.14
An example of a dataset with the cost of each arc for a shortest path problem.

Similar to Section 3.1, some parameters need to be set before calling the %orshortpath macro (See program "sasor_3_3.sas").

```
* SAS macro for shortest path problem;
%let _title = 'Example 3.3: An example of a shortest path
network.';
%let _data = 'c:\sasor\data3_3.txt';
%let _sourcenode = 'City1';
%let _sinknode = 'City7';
%let _cost = cost;
%let _tail = origin;
%let _head = dest;
%orshortpath;
```

This code determines the results based on the specified parameters and the cost of each arc saved in the text file; it also produces a macro variable (_ORNETFL) at termination. The SAS code for the shortest path problem contains three macros: data-handling (%data), model-building (%model), and report-writing (%report).

```
* The data macro;
%macro data;
* Import text tab delimited data file to SAS data file;
 proc import
        datafile = &_data
        out = dpath
        dbms = tab
        replace;
        getnames = yes;
 run;
%mend data;
```

```
* The model-building macro;
%macro model;
 proc netflow
        shortpath
        sourcenode = &_sourcenode
        sinknode = &_sinknode
        arcdata = dpath
        arcout = arcout1;
        cost &_cost;
        tail &_tail;
        head &_head;
 run;
%put &_ORNETFL;
%mend model;
```

```
* The report-writing macro;
%macro report;
 title &_title ', how to get from ' &_sourcenode ' to '
&_sinknode ;
 * Sort results by origin;
 proc sort
       data = arcout1
       out = result1;
       by _fcost_;
 run;
 * Print results sorted by origin;
 proc print
       data = result1;
       sum _fcost_;
run;
 * Sort results by total destination;
 proc sort
       data = arcout1
       out = result2;
       by _anumb_;
       run;
 * Print results sorted by total destination;
 proc print
       data = result2 (where = (_fcost_ ne 0));
       sum _fcost_;
 run;
%mend report;
```

```
%macro orshortpath;
 %data;
 %model;
 %report;
%mend orshortpath;
```

3.3.4 Sample Results from ORSHORTPATH Macro: Output from SAS

Figure 3.15 shows all the information concerning 10 arcs, including the unit cost of each arc, the upper and lower bounds of each arc, the supply of each tail or origin, the demand of each head or destination, number of flows in each arc, and the cost of flows in each arc. Figure 3.16 shows the same information as in Figure 3.15, except that those arcs having a unit of flow are shown only. According to Figure 3.16, the shortest path is 1 – 2 – 4 – 7, and the optimal solution value is $19. As shown in Figure 3.17, the total number of iterations is four.

Output - (Untitled)

Obs	Origin	Dest	cost	CAPAC	_LO_	_SUPPLY_	_DEMAND_	FLOW	FCOST	RCOST	ANUMB	TNUMB	_STATUS_	
1	City1	City3	10	99999999	0	1	.	0	0	.	2	1	KEY_ARC	BASIC
2	City2	City3	1	99999999	0	.	.	0	0	0	3	2	LOWERBD	NONBASIC
3	City2	City5	5	99999999	0	.	.	0	0	2	5	2	LOWERBD	NONBASIC
4	City6	City5	2	99999999	0	.	.	0	0	2	6	6	LOWERBD	NONBASIC
5	City3	City6	3	99999999	0	.	.	0	0	1	7	3	LOWERBD	NONBASIC
6	City5	City7	8	99999999	0	.	1	0	0	1	9	5	LOWERBD	NONBASIC
7	City6	City7	7	99999999	0	.	1	0	0	.	10	6	KEY_ARC	BASIC
8	City4	City7	4	99999999	0	.	1	1	4	.	8	4	KEY_ARC	BASIC
9	City2	City4	6	99999999	0	.	.	1	6	.	4	2	KEY_ARC	BASIC
10	City1	City2	9	99999999	0	1	.	1	9	.	1	1	KEY_ARC	BASIC
									== 19					

FIGURE 3.15
Result of %orshortpath.

Output - (Untitled)

Obs	Origin	Dest	cost	CAPAC	_LO_	_SUPPLY_	_DEMAND_	FLOW	FCOST	RCOST	ANUMB	TNUMB	_STATUS_	
1	City1	City2	9	99999999	0	1	.	1	9	.	1	1	KEY_ARC	BASIC
4	City2	City4	6	99999999	0	.	.	1	6	.	4	2	KEY_ARC	BASIC
8	City4	City7	4	99999999	0	.	1	1	4	.	8	4	KEY_ARC	BASIC
									== 19					

FIGURE 3.16
Result of %orshortpath.

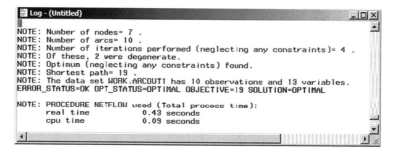

```
NOTE: Number of nodes= 7 .
NOTE: Number of arcs= 10 .
NOTE: Number of iterations performed (neglecting any constraints)= 4 .
NOTE: Of these, 2 were degenerate.
NOTE: Optimum (neglecting any constraints) found.
NOTE: Shortest path= 19 .
NOTE: The data set WORK.ARCOUT1 has 10 observations and 13 variables.
ERROR_STATUS=OK OPT_STATUS=OPTIMAL OBJECTIVE=19 SOLUTION=OPTIMAL

NOTE: PROCEDURE NETFLOW used (Total process time):
      real time          0.43 seconds
      cpu time           0.09 seconds
```

FIGURE 3.17
Log for %orshortpath.

3.3.5 Exercise

Use the codes developed in this chapter and solve the following shortest path problem found in Table 3.10.

Solution:

- Create the data in a text file (see "data3_3_exercise.txt").
- Run the following code (see program "sasor_3_3_ exercise.sas"):

TABLE 3.10

A Shortest Path Exercise

Origin i	Destination j						s_i
	2	**3**	**4**	**5**	**6**	**7**	
1	1 x_{12}	7 x_{13}	∞ x_{14}	∞ x_{15}	∞ x_{16}	∞ x_{17}	1
2	0 x_{22}	9 x_{23}	10 x_{24}	2 x_{25}	∞ x_{26}	∞ x_{27}	1
3	∞ x_{32}	0 x_{33}	∞ x_{34}	∞ x_{35}	6 x_{36}	∞ x_{37}	1
4	∞ x_{42}	∞ x_{43}	0 x_{44}	∞ x_{45}	∞ x_{46}	4 x_{47}	1
5	∞ x_{52}	∞ x_{53}	∞ x_{54}	0 x_{55}	∞ x_{56}	8 x_{57}	1
6	∞ x_{62}	∞ x_{63}	∞ x_{64}	5 x_{65}	0 x_{66}	3 x_{67}	1
dj	1	1	1	1	1	1	

```
* SAS macro for shortest path problem: solution to exercise
3.3.;
%let _title = 'An example of a shortest path network,
solution to exercise 3.3.';
%let _data = 'c:\sasor\data3_3_exercise.txt';
%let _sourcenode = 'City1';
%let _sinknode = 'City7';
%let _cost = cost;
%let _tail = origin;
%let _head = dest;
option nodate;
%orshortpath;
```

The following solution is given by SAS:

4

Project Scheduling

In this chapter, we focus on two project-scheduling techniques and demonstrate how SAS/OR® can be applied to solve these problems to optimality. The techniques include critical path analysis and program evaluation and review technique.

4.1 Critical Path Analysis

4.1.1 Concept of Critical Path Analysis

Critical path analysis is a network-based method designed to aid in scheduling, monitoring, and controlling large and complex projects, particularly in the construction industry. A project can be defined as a series of related activities or tasks, with each activity consuming time and resources. Finding a critical path is a major part of project management. The activities on the critical path represent tasks that will delay the entire project if they are not completed on time. Based on the critical path analysis, project managers can reschedule and reallocate labor and financial resources so that the critical tasks can be completed on time. Critical path analysis is important because it can answer a number of questions about projects, such as:

1. When will the entire project be finished?
2. What are the critical activities or tasks in the project?
3. Which activities or tasks can be delayed if necessary and by how long without delaying the entire project?

Linear programming (LP) can be used to formulate such a problem and then yield the critical path. Consider a project with n activities. The objective is to minimize the time required to complete the entire project. For each activity, we are certain that before node j occurs, node i must occur and activity on arc (i, j) must be completed. The time required by an activity on arc (i, j) is denoted as t_{ij}. By introducing decision variables x_j to represent the completion time of an activity on arc (i, j), the mathematical model for the critical path analysis can be written as shown in Model 4.1.1.

Model 4.1.1 Standard critical path analysis model

$$\text{Minimize } z = x_n - x_o \qquad (4.1.1)$$

subject to

$$x_j \geq x_i + t_{ij} \quad i,j = 0,1,\dots,n \qquad (4.1.2)$$

$$\text{All } x_j \geq 0$$

Model 4.1.1 is referred to as the *critical path analysis model*. Objective function 4.1.1 minimizes the time required to complete the entire project. Constraint set 4.1.2 is a precedence relationship that guarantees that an activity cannot be performed until its immediate predecessors are completed.

4.1.2 Example of Critical Path Analysis

Before identifying the critical path, there are four steps to follow:

1. Define the project and its activities.
2. Define the precedence relationships among the activities.
3. Assign the time requirement to each activity.
4. Draw the network connecting all of the activities.

Table 4.1 shows a project with eight activities in which activities A and B are done first because they have no predecessor and activity H is the terminal point. The precedence relationships among the activities, as well as their time requirements, are listed in the table. Table 4.2 shows the same information as in Table 4.1, except that the immediate successors of activities are shown.

After defining the precedence relationships among the activities and the time requirements to each activity, a network representing the project can be constructed (Figure 4.1).

As shown in Figure 4.1, each activity is represented by a directional arc or arrow. This type of project network is regarded as activity-on-arc (AOA) network. There are two crucial rules for the construction of AOA network: (1) each activity is represented by exactly one arc or arrow in the network, and (2) each activity must be identified by two nodes. For example, activity A, which requires 2 units of time to complete, is linked by two nodes. Node 0 is the starting point of activity A, and node 1 is the terminal point of activity A. To prevent a violation of the rules, it is sometimes necessary to use a dummy activity with 0 task time in the network. For example, activities A and B are predecessors of activity D. In this case, we can add a dummy activity, shown by a dashed arrow, pointing from node 1 to 2. By adding this dummy

TABLE 4.1

Precedence Relationships and Time
Requirements of a Project

Activity	Immediate Predecessors	Time
A	–	2
B	–	3
C	A	2
D	A, B	4
E	C	4
F	C	3
G	D, E	5
H	F, G	2

TABLE 4.2

Precedence Relationships and Time
Requirements of a Project

Activity	Immediate Successors	Time
A	C, D	2
B	D	3
C	E, F	2
D	G	4
E	G	4
F	H	3
G	H	5
H	–	2

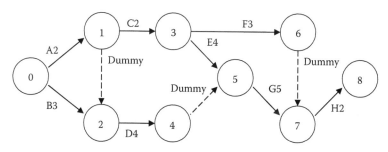

FIGURE 4.1
A project network.

activity, we are able to illustrate how activities A and B are both predecessors of activity D. Activity D cannot be performed until activities A and B are both completed.

By introducing decision variables x_j to represent the completion time of an activity on arc (i, j), this critical path problem can be formulated as shown in Model 4.1.2.

Model 4.1.2 Example of formulation of critical path analysis problem

$$\text{Minimize } x_8 - x_0 \qquad (4.1.3)$$

subject to

$$x_1 - x_0 \geq 2 \qquad (4.1.4)$$

$$x_2 - x_0 \geq 3 \qquad (4.1.5)$$

$$x_3 - x_1 \geq 2 \qquad (4.1.6)$$

$$x_4 - x_2 \geq 4 \qquad (4.1.7)$$

$$x_5 - x_3 \geq 4 \qquad (4.1.8)$$

$$x_6 - x_3 \geq 3 \qquad (4.1.9)$$

$$x_7 - x_5 \geq 5 \qquad (4.1.10)$$

$$x_8 - x_7 \geq 2 \qquad (4.1.11)$$

$$x_2 - x_1 \geq 0 \qquad (4.1.12)$$

$$x_5 - x_4 \geq 0 \qquad (4.1.13)$$

$$x_7 - x_6 \geq 0 \qquad (4.1.14)$$

$$\text{All } x_j \geq 0$$

Constraint sets 4.1.4 to 4.1.11 are the precedence relationship constraints for the actual activities represented by the solid arrows. For example, constraint set 4.1.6 ensures that activity C on arc (1, 3) cannot start until activity A on arc (0, 1) is completed. Constraint sets 4.1.12 to 4.1.14 are the precedence relationship constraints for the dummy activities represented by the dashed arrows. Because a dummy activity has no duration, the right-side values of the last three constraint sets are 0. If all the right-side values of the constraint sets are integral, we can be sure that the optimal solution to Model 4.1.2 will be integral, even when we solve the problem as an LP problem.

4.1.3 ORCPM: SAS Code for Critical Path Analysis

ORCPM is a macro that finds the critical path, which is very useful in project management (see program "sasor_4_1.sas"). The primary procedure used for

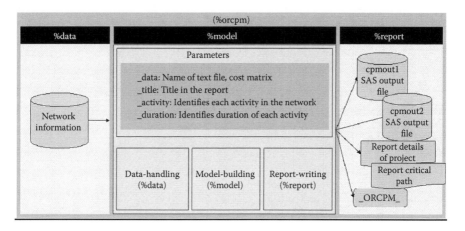

FIGURE 4.2
Data flow in ORCPM.

a critical path analysis problem is PROC CPM. A full syntax of this procedure is available in Appendix 5.

Figure 4.2 illustrates the data flow in ORCPM. It shows:

- The dataset that is required for ORCPM, in which the tasks, duration of each task, and succeeded task following each task are specified
- The macros (%data, %model, and %report)
- The results datasets that are available for print or can be used for further analysis

In the rest of this section, the procedure for solving a critical path problem (ORCPM) in SAS, together with an example, is explained. ORCPM runs three macros: data-handling (%data), model-building (%model), and report-writing (%report).

4.1.4 ORCPM: Data-Handling Macro (%data)

This part of ORCPM processes the data into a format that is suitable for PROC CPM. ORCPM requires one dataset containing the names and duration of the tasks and the names of the successor tasks. The data set should be a .txt file, which is saved as "text tab delimited." The activities' names must start with a letter and may contain up to 50 characters. The variable names must be listed in the first row of the data file. The task names should be listed in the first column; the task names and duration variables used in the dataset should be given to the macro before calling it. An example of data file is seen in Figure 4.3.

One parameter needs to be set before calling the data macro:

_data: Indicates the name and location of the data file (a text tab delimited file) and contains the cost matrix

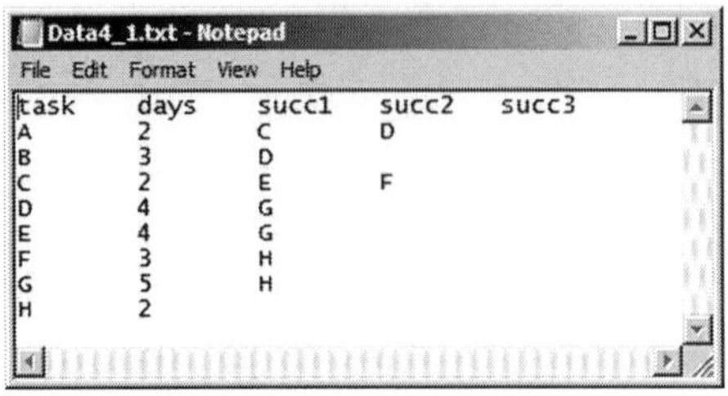

Data4_1.txt - Notepad

File Edit Format View Help

task	days	succ1	succ2	succ3
A	2	C	D	
B	3	D		
C	2	E	F	
D	4	G		
E	4	G		
F	3	H		
G	5	H		
H	2			

FIGURE 4.3
An example of data set, activity network.

```
* The data-handling macro;
%macro data;
* Import text tab delimited data file to SAS data file;
 proc import
  datafile = &dcpm
  out = dcpm
  dbms = tab
  replace;
  getnames = yes;
  run;
%mend data;
```

4.1.5 ORCPM: Model-Building Macro (%model)

This part of ORCPM calls PROC CPM to solve the network. Two parameters need to be set before calling the model macro:

_activity: Identifies each activity in the network

_duration: Identifies the duration of each activity

The SAS macro for model-building is as follows

```
* The model-building macro;
%macro model;
 proc cpm
      out = cpmout1;
      activity &_activity;
      duration &_duration;
      successor succ1 succ2 succ3;
 run;
 %put &_ORCPM_;
%mend model;
```

4.1.6 ORCPM: Report-Writing Macro (%report)

The outputs from ORCPM include a report that contains all information concerning project management, including earliest start, earliest finish, latest start, and latest finish times for each activity as well as report on critical path analysis. This information is saved in the "cpmout1" and "cpmout2" datasets. One parameter needs to be set before calling this macro:

_title: Gives a title in the output of the SAS

```
%macro report;
  data cpmout (where = (T_float = 0));
   set cpmout1;
  run;

  title &_title;
  proc print
      data = cpmout1 ;
  run;

  title &_title;
  proc print
      data = cpmout ;
      sum &_duration;
  run;
%mend report;
```

4.1.7 ORCPM: Macro (%orcpm)

To make the system as user friendly as possible, the %orcpm macro combines the data-handling, model-building, and report-writing codes.

```
* Invoke PROC CPM to schedule the project specifying
* the ACTIVITY, DURATION, and SUCCESSOR variables;
%macro orcpm;
  %data;
  %model;
  %report;
%mend orcpm;
```

In this code, the %orcpm macro is used to manage the codes explained earlier, including data-handling, model-building, and report-writing. To get the results, the user needs to set up the parameters and run only one statement:

```
%ORCPM;
```

Output - (Untitled)

Obs	task	succ1	succ2	succ3	days	E_START	E_FINISH	L_START	L_FINISH	T_FLOAT	F_FLOAT
1	A	C	D		2	0	2	0	2	0	0
2	B	D			3	0	3	1	4	1	0
3	C	E	F		2	2	4	2	4	0	0
4	D	G			4	3	7	4	8	1	1
5	E	G			4	4	8	4	8	0	0
6	F	H			3	4	7	10	13	6	6
7	G	H			5	8	13	8	13	0	0
8	H				2	13	15	13	15	0	0

FIGURE 4.4
Result of %orcpm, details of the project.

Output - (Untitled)

Obs	task	succ1	succ2	succ3	days	E_START	E_FINISH	L_START	L_FINISH	T_FLOAT	F_FLOAT
1	A	C	D		2	0	2	0	2	0	0
2	C	E	F		2	2	4	2	4	0	0
3	E	G			4	4	8	4	8	0	0
4	G	H			5	8	13	8	13	0	0
5	H				2	13	15	13	15	0	0
					===========						
					15						

FIGURE 4.5
Result of %orcpm, details of the critical path.

4.1.8 Instructions for Using ORCPM Macro

This section presents SAS code for the earlier example of a critical path analysis problem with eight activities, as shown in Table 4.2. The data are saved in file "data4_1.txt."

A user needs to set the parameters as required and run the following code:

```
%let _title = 'Example 4.1. A pro ject network: Activity-
On-Node Format : Critical Path Analysis';
%let dcpm = 'c:/sasor/Data4_1.txt';
%let _activity = task;
%let _duration = days;
%orcpm;
```

This code determines the results based on the specified parameters and the network information saved in the text file and also produces a macro variable (_ORCPM_) at termination. The user can examine the results of this macro variable, examine whether PROC CPM ran correctly, and examine what error or difficulty it encountered. Because _ORCPM_ is a standard SAS macro variable, it can be used as all macro variables can be used.

4.1.9 Sample Results from ORCPM Macro: Output from SAS

The results of running this code are presented in Figures 4.4 and 4.5. Figure 4.4 shows all the information concerning eight activities, including

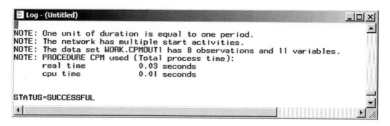

FIGURE 4.6
Log for %orcpm.

TABLE 4.3

Precedence Relationships and Time
Requirements of a Project

Activity	Immediate Successor(s)	Time
A	B, C, D	1
B	E	4
C	F	3
D	F, H	7
E	G	6
F	G	2
G	I	7
H	I	9
I	–	4

successors of each task, time requirements of each task, and four times for each task (i.e., earliest start, earliest finish, latest start, and latest finish times). Figure 4.5 shows the same information as Figure 4.4, except that those tasks on the critical path are shown only. According to Figure 4.5, the critical path is A – C – E – G – H and the optimal solution value is 15 days.

The printout of the log file (Figure 4.6) shows that the CPM procedure terminated successfully. There are 8 observations and 11 variables in the output dataset.

4.1.10 Exercise

Use the codes developed in this chapter and solve the critical path problem shown in Table 4.3.

Solution:

- Create the data in a text file (see " data4_1_exercise.txt").
- Run the following code (see program "sasor_4_1_ exercise.sas"):

```
* An SAS procedure for project network of type Activity-On-
Node: solution to exercise 4.1.;
```

```
%let _title = 'A project network: Activity-On-Node Format:
Critical Path, Analysis, solution to exercise 4.1.';
%let dcpm = 'c:/sasor/Data4_1_exercise.txt';
%let _activity = task;
%let _duration = days;
%orcpm;
```

The following solution is given by SAS:

VIEWTABLE: Work.Cpmout1

	task	succ1	succ2	succ3	days	Early Start	Early Finish	Late Start	Late Finish	Total Float	Free Float
1	A	C	D		2	0	2	0	2	0	0
2	B	D			3	0	3	1	4	1	0
3	C	E	F		2	2	4	2	4	0	0
4	D	G			4	3	7	4	8	1	1
5	E	G			4	4	8	4	8	0	0
6	F	H			3	4	7	10	13	6	6
7	G	H			5	8	13	8	13	0	0
8	H				2	13	15	13	15	0	0

VIEWTABLE: Work.Cpmout

	task	succ1	succ2	succ3	days	Early Start	Early Finish	Late Start	Late Finish	Total Float	Free Float
1	A	C	D		2	0	2	0	2	0	0
2	C	E	F		2	2	4	2	4	0	0
3	E	G			4	4	8	4	8	0	0
4	G	H			5	8	13	8	13	0	0
5	H				2	13	15	13	15	0	0

4.2 Program Evaluation and Review Technique (PERT)

4.2.1 Concept of PERT

As with the critical path analysis discussed in Section 4.1, the *program evaluation and review technique (PERT)* is a network-based method designed to manage large and complex projects. PERT is an extension of critical path analysis because it not only can answer the three important questions discussed in Section 4.1.1, but it can also answer a fourth critical question: what is the probability that the project will be completed by the due date?

To answer these questions, there are four steps to follow:

1. Define the project and its activities.
2. Define the precedence relationships among the activities.
3. Assign the probabilistic time estimates to each activity.
4. Draw the network connecting all of the activities.

This procedure is almost the same as that of critical path analysis except that PERT employs three probabilistic time estimates for an activity in step 3

TABLE 4.4

Precedence Relationships and Time Requirements of a Project

Activity	Immediate Predecessor(s)	Time		
		Optimistic	Most Likely	Pessimistic
A	–	5	6	7
B	–	2	3	10
C	–	1	2	3
D	A	4	7	10
E	B	1	3	11
F	D, E	5	9	13
G	B, C	1	2	9
H	F, G	4	6	8

TABLE 4.5

Precedence Relationships and Time Requirements of a Project

Activity	Immediate Successor(s)	Time		
		Optimistic	Most Likely	Pessimistic
A	D	5	6	7
B	E, G	2	3	10
C	G	1	2	3
D	F	4	7	10
E	F	1	3	11
F	H	5	9	13
G	H	1	2	9
H	–	4	6	8

whereas there is only one time factor for each activity in the critical path analysis. For PERT, the time estimates are the optimistic time, a; the most likely time, m; and the pessimistic time, b. The optimistic time is the shortest possible time to complete an activity if everything goes as planned, the most likely time is the most realistic time to complete an activity if it was repeated many times, and the pessimistic time is the longest possible time to complete an activity that takes all unfavorable conditions into consideration.

4.2.2 Example of PERT

Table 4.4 lists the number of activities in the project, the precedence relationships among the activities, and the three time estimates for every activity. Table 4.5 shows the same information as Table 4.4, except that the immediate successors of activities are shown.

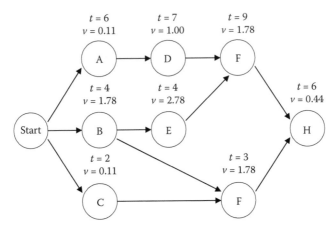

FIGURE 4.7
A project network.

The expected activity time, t, and the variance of activity completion time, v, are computed using equations 4.2.1 and 4.2.2, respectively. In addition, the project completion probability is computed using equation 4.2.3, where Z is the number of standard deviations the due date lies from the expected project completion date, X is the due date, T is the expected project completion date, and V is the variance of project completion time. Figure 4.7 shows the project network.

$$t = \frac{a + 4m + b}{6} \tag{4.2.1}$$

$$v = \left(\frac{b - a}{6}\right)^2 \tag{4.2.2}$$

$$Z = \frac{X - T}{\sqrt{v}} \tag{4.2.3}$$

4.2.3 ORPERT: SAS Code for PERT

PROC CPM, explained in Section 4.1, is a macro that can also be used to solve PERT, or project management with uncertainty, which is a special case of the critical path analysis problem. Hence, we use a similar macro with some minor changes to solve this. The new macro is called ORPERT (see program "sasor_4_2.sas"). The only difference is that the dataset contains three further variables, including "opti", "most", and "pessi" for the optimistic time, most likely time, and pessimistic time, respectively, in the project. An example dataset is shown in Figure 4.8.

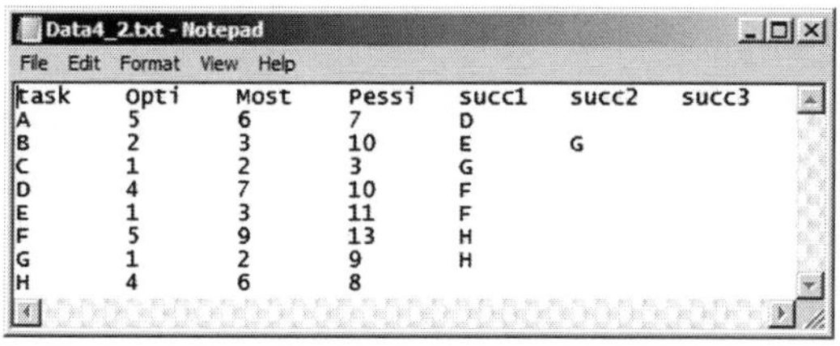

task	Opti	Most	Pessi	succ1	succ2	succ3
A	5	6	7	D		
B	2	3	10	E	G	
C	1	2	3	G		
D	4	7	10	F		
E	1	3	11	F		
F	5	9	13	H		
G	1	2	9	H		
H	4	6	8			

FIGURE 4.8
An example dataset, details of the project with activities and duration of each activity.

Similar to Section 4.1, some parameters need to be set before calling the %orpert macro:

```
%let _title='Example 4.2. A project network: Activity-On-
Node Format: employs three probabilistic time estimates for
each activity';
%let _data = 'c:/sasor/Data4_2.txt';
%let _activity = task;
%let _duration = days;
%orpert;
```

This code determines the results based on the specified activities and the estimated duration for each activity as saved in the text file; it also produces a macro variable (_ORCPM_) at termination. The complete SAS code for PERT, or critical path analysis with uncertainty, contains three macros (%data, %model, and %report).

```
/* Activity-on-Node representation of the project with
uncertainty */

* The data-handling macro;
%macro data;
* Import text tab delimited data file to SAS data file;
 proc import
       datafile = &_data
       out = dpert
       dbms = tab
       replace;
       getnames = yes;
 run;

  data dpert;
  set dpert;
  days = (Opti + 4*most + pessi)/6;
  variance = ((Opti-pessi)/6)**2;
 run;
%mend data;
```

```
* The model-building macro;
%macro model;
 proc cpm
        out = pertout;
        activity &_activity;
        duration &_duration;
        successor succ1 succ2 succ3;
 run;
 %put &_ORCPM_;
%mend model;
```

```
* The report-writing macro;
%macro report;

 data pertout1 ;
  merge dpert pertout;
  by task;
 run;
 data pertout2 (where = (T_float = 0));
  merge dpert pertout;
  by task;
 run;

 title &_title;
 proc print
     data = pertout1 ;
 run;

 proc print
     data = pertout2 ;
     sum T_float;
     sum days;
     sum variance;
 run;
%mend report;
```

```
%macro orpert;
 %data;
 %model;
 %report;
%mend orpert;
```

4.2.4 Sample Results from ORPERT Macro: Output from SAS

The results of running this code are presented in Figures 4.9 and 4.10. Figure 4.9 shows all information concerning the eight tasks or activities, including;

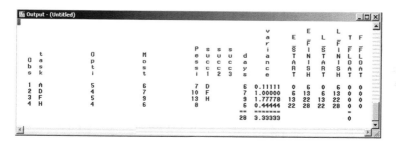

	task	succ1	succ2	succ3	days	Early Start	Early Finish	Late Start	Late Finish	Total Float	Free Float
1	A	D			6	0	6	0	6	0	0
2	B	E	G		4	0	4	5	9	5	0
3	C	G			2	0	2	17	19	17	2
4	D	F			7	6	13	6	13	0	0
5	E	F			4	4	8	9	13	5	5
6	F	H			9	13	22	13	22	0	0
7	G	H			3	4	7	19	22	15	15
8	H				6	22	28	22	28	0	0

FIGURE 4.9
Result of %orpert, all activities.

FIGURE 4.10
Result of %orpert, activities on the critical path.

- The three time estimates for each task (i.e., optimistic time, most likely time, and pessimistic time)
- The successors to each task
- The time requirement of each task
- The variance of each task
- The four times of each task (i.e., earliest start, earliest finish, latest start, and latest finish times)

Figure 4.10 shows the same information as Figure 4.9, except that only those tasks on the critical path are shown. According to Figure 4.10, the critical path is A – D – F – H. The optimal solution value, or the expected project completion time, is 28 days. The project variance is 3.33 days.

Assume that the due date for this example is on day 31 and that the project is started on day 1. The project manager would like to know the probability that the project will be completed on or before day 31. To achieve this goal, the Z value is computed using equation 4.2.3—that is, $Z = (31 - 28)/\sqrt{3.33} = 1.64$.

According to the cumulative standard normal distribution table as shown in Appendix 6, the probability that the project will be completed on or before day 31 is 0.9495, or 94.95%.

The printout of the log file (Figure 4.11) shows that the PERT procedure terminated successfully. There are 8 observations and 11 variables in the output dataset.

FIGURE 4.11

Log for %orpert.

TABLE 4.6

Precedence Relationships and Time Requirements of a Project

Activity	Immediate Successor(s)	Time		
		Optimistic	Most Likely	Pessimistic
A	B, C, D	0.5	1	1.5
B	E	2	3	10
C	F	1	2	9
D	F, H	4	7	10
E	G	5	6	7
F	G	1	2	3
G	I	4	7	10
H	I	5	9	13
I	–	1	3	11

4.2.5 Exercise

Use the codes developed in this chapter and solve the PERT problem in Table 4.6

Solution:

- Create the data in a text file (see "data4_2_exercise.txt").
- Run the following code (see program "sasor_4_2_ exercise.sas"):

```
* An SAS procedure for project network of type Activity-On-
Node: solution to exercise 4.2.;
%let _title = 'A project network: Activity-On-Node Format:
solution to exercise 4.2.';
%let dcpm = 'c:/sasor/Data4_2_exercise.txt';
%let _activity = task;
%let _duration = days;
%orpert;
```

The following solution is given by SAS:

VIEWTABLE: Work.Pertout1

task	Opti	Most	Pessi	succ1	succ2	succ3	days	variance	Early Start	Early Finish	Late Start	Late Finish	Total Float	Free Float
A	0.5	1	1.5	B	C	D	1	0.0277777778	0	1	0	1	0	0
B	2	3	10	E			4	1.7777777778	1	5	1	5	0	0
C	1	2	9	F			3	1.7777777778	1	4	6	9	5	4
D	4	7	10	F	H		7	1	1	8	2	9	1	0
E	5	6	7	G			6	0.1111111111	5	11	5	11	0	0
F	1	2	3	G			2	0.1111111111	8	10	9	11	1	1
G	4	7	10	I			7	1	11	18	11	18	0	0
H	5	9	13	I			9	1.7777777778	8	17	9	18	1	1
I	1	3	11				4	2.7777777778	18	22	18	22	0	0

VIEWTABLE: Work.Pertout2

task	Opti	Most	Pessi	succ1	succ2	succ3	days	variance	Early Start	Early Finish	Late Start	Late Finish	Total Float	Free Float
A	0.5	1	1.5	B	C	D	1	0.0277778	0	1	0	1	0	0
B	2	3	10	E			4	1.7777778	1	5	1	5	0	0
E	5	6	7	G			6	0.1111111	5	11	5	11	0	0
G	4	7	10	I			7	1	11	18	11	18	0	0
I	1	3	11				4	2.7777778	18	22	18	22	0	0

Output - (Untitled)

```
                                         v           E       L     L
                                         a       E   F   L   F   T  F
       t              O          M        r       a   S   F   S   F   F  F
 O     a              p          o    P  s  s  s  i    S   T   I   T   I  L  L
 b     s              t          o    e  u  u  u  d   a   T   N   T   N   L  L
 s     k              i          s    s  c  c  c  a   n   A   I   A   I   0  0
                                 t    i  1  2  3  y   c   R   S   R   S   A  A
                                         s           e   T   H   T   H   T  T

 1  A          0.5           1    1.5 B  C  D  1  0.02778  0   1   0   1  0  0
 2  B            2           3     10 E        4  1.77778  1   5   1   5  0  0
 3  E            5           6      7 G        6  0.11111  5  11   5  11  0  0
 4  G            4           7     10 I        7  1.00000 11  18  11  18  0  0
 5  I            1           3     11          4  2.77778 18  22  18  22  0  0
                                                 ==  =======                  =
                                                 22  5.69444                  0
```

5

Layout Decision

In this chapter, we present the assembly line balancing problem and demonstrate how SAS/OR® can be applied to solve the problem to optimality.

5.1 Assembly Line–Balancing Problem

5.1.1 Concept of Assembly Line–Balancing Problem

Assembly line balancing is a product-oriented layout technique in operations management. Product-oriented layouts are designed for high-volume, low-variety products or continuous production. The problem of assembly line balancing is how to assign tasks to workstations while meeting production requirements at a minimum imbalance between labors or machines. Minimization of imbalance leads to minimization of idle time along the assembly line and maximization utilization of labors and machines.

Integer linear programming (ILP) can be used to formulate an assembly line balancing problem and then yield the optimal product layout. Consider a job with a series of tasks ($i = 1, 2, \ldots, m$) to be assigned to a certain number of workstations ($j = 1, 2, \ldots, n$). The objective is to minimize the number of workstations, A_j, used to complete the work. Given the precedence relationships, the earliest workstation to which task i can be assigned is denoted as E_i, and the latest workstation to which task i can be assigned is denoted as L_i. The time required by task i is t_i, and the theoretical cycle time is C. W_j is the subset of all tasks that can be assigned to workstation j, and $||W_j||$ is the number of tasks in subset W_j. P_i is the set of tasks that must proceed task i, and S_i is the set of tasks that must succeed task i. By introducing decision variables x_{ij} to represent the assignment of task i to workstation j, the mathematical model for the assembly line balancing problem can be written as shown in Model 5.1.1 (Patterson and Albracht 1975; Gökçen and Erel 1998; Ağpak and Gökçen 2005).

Model 5.1.1 Standard assembly line balancing model

$$\text{Minimize } z = \sum_{j=1}^{n} A_j \qquad (5.1.1)$$

Subject to

$$\sum_{j=E_i}^{L_i} x_{ij} = 1 \quad i = 1, 2, \ldots, m \tag{5.1.2}$$

$$\sum_{i \in W_j} t_i x_{ij} \leq C \quad j = 1, 2, \ldots, n \tag{5.1.3}$$

$$\sum_{j=E_a}^{L_a} j x_{aj} \leq \sum_{j=E_b}^{L_b} j x_{bj} \quad \text{for } \forall (a, b) \in P \tag{5.1.4}$$

$$\sum_{i \in W_j} x_{ij} \leq \|W_j\| A_j \quad j = 1, 2, \ldots, n \tag{5.1.5}$$

$$A_j \geq A_{j+1} \quad j = 1, 2, \ldots, n \tag{5.1.6}$$

All x_{ij} and $A_j = 0$ or 1

where

$$E_i = \left[\left(t_i + \sum_{k \in P_i} t_k \right) / C \right]^+ \tag{5.1.7}$$

$$L_i = n + 1 - \left[\left(t_i + \sum_{k \in S_i} t_k \right) / C \right]^+ \tag{5.1.8}$$

Model 5.1.1 is referred to as the *assembly line–balancing model*. Objective function 5.1.1 minimizes the number of workstations used to complete the work. Constraint set 5.1.2 is an assignment constraint that guarantees that each task is assigned to exactly one workstation. Constraint set 5.1.3 is a cycle time constraint that ensures that the sum of task times assigned to each workstation does not exceed the theoretical cycle time (i.e., maximum time allowed at each workstation). Constraint set 5.1.4 is a precedence relationship constraint that guarantees that task b cannot be performed until task a is finished when task a is an immediate predecessor of task b. Constraint set 5.1.5 is a workstation constraint that ensures that a workstation is used if the task is assigned to it. Constraint set 5.1.6 ensures that an earlier workstation is used if the task is assigned to the workstation that proceeds it. In both equations 5.1.7 and 5.1.8, $[y]^+$ denotes the smallest integer greater than or equal to y.

Before identifying the optimal product layout, there are four steps to follow:

1. Define the work and its tasks.
2. Define the precedence relationships among the tasks.

3. Assign the time requirement to each task.

4. Draw the network connecting all the tasks.

5.1.2 Example of Assembly Line–Balancing Problem

Table 5.1 shows a job with eight tasks in which task 1 is done first because it has no predecessor and task 8 is the terminal point. The precedence relationships among the tasks and the time requirements of the tasks are listed in Table 5.1.

After defining the precedence relationships among the tasks and the time requirements for each task, a network representing the work can be constructed (Figure 5.1).

As shown in Figure 5.1, each task is represented by a node instead of a directional arc. This type of network is regarded as an activity-on-node (AON) network. The time requirement of each task, t_i, is shown above the

TABLE 5.1

Precedence Relationships and
Time Requirements of a Work

Task	Immediate Predecessors	Time
1	–	1.0
2	1	0.6
3	1	1.6
4	2	2.7
5	2	1.2
6	4	0.9
7	5	3.4
8	3, 6, 7	1.1

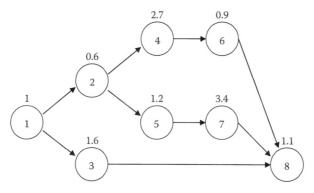

FIGURE 5.1
An assembly line–balancing network.

TABLE 5.2

Earliest and Latest
Workstations to Which Tasks
Can be Assigned

Task	E_i	L_i
1	1	2
2	1	3
3	1	5
4	2	4
5	1	4
6	2	5
7	2	4
8	4	5

node. It is assumed that the theoretical cycle time, C, is 3.7 units of time. The earliest workstation, E_i, and the latest workstation, L_i, to which task i can be assigned are listed in Table 5.2. Note that the maximum number of workstations available is five (i.e., $n = 5$).

By introducing decision variables x_{ij} to represent the assignment of task i to workstation j, this assembly line balancing problem can be formulated as:

$$\text{Minimize } z = A_1 + A_2 + A_3 + A_4 + A_5 \tag{5.1.9}$$

subject to

$$x_{11} + x_{12} = 1 \tag{5.1.10}$$

$$x_{21} + x_{22} + x_{23} = 1 \tag{5.1.11}$$

$$x_{31} + x_{32} + x_{33} + x_{34} + x_{35} = 1 \tag{5.1.12}$$

$$x_{42} + x_{43} + x_{44} = 1 \tag{5.1.13}$$

$$x_{51} + x_{52} + x_{53} + x_{54} = 1 \tag{5.1.14}$$

$$x_{62} + x_{63} + x_{64} + x_{65} = 1 \tag{5.1.15}$$

$$x_{72} + x_{73} + x_{74} = 1 \tag{5.1.16}$$

$$x_{84} + x_{85} = 1 \tag{5.1.17}$$

$$x_{11} + 0.6\, x_{21} + 1.6\, x_{31} + 1.2\, x_{51} \leq 3.7 \tag{5.1.18}$$

$$x_{12} + 0.6\ x_{22} + 1.6\ x_{32} + 2.7\ x_{42} + 1.2\ x_{52} + 0.9\ x_{62} + 3.4\ x_{72} \leq 3.7 \quad (5.1.19)$$

$$0.6\ x_{23} + 1.6\ x_{33} + 2.7\ x_{43} + 1.2\ x_{53} + 0.9\ x_{63} + 3.4\ x_{73} \leq 3.7 \quad (5.1.20)$$

$$1.6\ x_{34} + 2.7\ x_{44} + 1.2\ x_{54} + 0.9\ x_{64} + 3.4\ x_{74} + 1.1\ x_{84} \leq 3.7 \quad (5.1.21)$$

$$1.6\ x_{35} + 0.9\ x_{65} + 1.1\ x_{85} \leq 3.7 \quad (5.1.22)$$

$$x_{11} + 2\ x_{12} - x_{21} - 2\ x_{22} - 3\ x_{23} \leq 0 \quad (5.1.23)$$

$$x_{11} + 2\ x_{12} - x_{31} - 2\ x_{32} - 3\ x_{33} - 4\ x_{34} - 5\ x_{35} \leq 0 \quad (5.1.24)$$

$$x_{21} + 2\ x_{22} + 3\ x_{23} - 2\ x_{42} - 3\ x_{43} - 4\ x_{44} \leq 0 \quad (5.1.25)$$

$$x_{21} + 2\ x_{22} + 3\ x_{23} - x_{51} - 2\ x_{52} - 3\ x_{53} - 4\ x_{54} \leq 0 \quad (5.1.26)$$

$$x_{31} + 2\ x_{32} + 3\ x_{33} + 4\ x_{34} + 5\ x_{35} - 4\ x_{84} - 5\ x_{85} \leq 0 \quad (5.1.27)$$

$$2\ x_{42} + 3\ x_{43} + 4\ x_{44} - 2\ x_{62} - 3\ x_{63} - 4\ x_{64} - 5\ x_{65} \leq 0 \quad (5.1.28)$$

$$x_{51} + 2\ x_{52} + 3\ x_{53} + 4\ x_{54} - 2\ x_{72} - 3\ x_{73} - 4\ x_{74} \leq 0 \quad (5.1.29)$$

$$2\ x_{62} + 3\ x_{63} + 4\ x_{64} + 5\ x_{65} - 4\ x_{84} - 5\ x_{85} \leq 0 \quad (5.1.30)$$

$$2\ x_{72} + 3\ x_{73} + 4\ x_{74} - 4\ x_{84} - 5\ x_{85} \leq 0 \quad (5.1.31)$$

$$x_{11} + x_{21} + x_{31} + x_{51} - 4\ A_1 \leq 0 \quad (5.1.32)$$

$$x_{12} + x_{22} + x_{32} + x_{42} + x_{52} + x_{62} + x_{72} - 7\ A_2 \leq 0 \quad (5.1.33)$$

$$x_{23} + x_{33} + x_{43} + x_{53} + x_{63} + x_{73} - 6\ A_3 \leq 0 \quad (5.1.34)$$

$$x_{34} + x_{44} + x_{54} + x_{64} + x_{74} + x_{84} - 6\ A_4 \leq 0 \quad (5.1.35)$$

$$x_{35} + x_{65} + x_{85} - 3\ A_5 \leq 0 \quad (5.1.36)$$

$$A_1 - A_2 \leq 0 \quad (5.1.37)$$

$$A_1 - A_3 \leq 0 \quad (5.1.38)$$

$$A_1 - A_4 \leq 0 \quad (5.1.39)$$

$$A_1 - A_5 \leq 0 \quad (5.1.40)$$

$$A_2 - A_3 \geq 0 \tag{5.1.41}$$

$$A_2 - A_4 \geq 0 \tag{5.1.42}$$

$$A_2 - A_5 \geq 0 \tag{5.1.43}$$

$$A_3 - A_4 \geq 0 \tag{5.1.44}$$

$$A_3 - A_5 \geq 0 \tag{5.1.45}$$

$$A_4 - A_5 \geq 0 \tag{5.1.46}$$

All X_{ij} and $A_J = 0$ or 1

Constraint sets 5.1.10 to 5.1.17 are the assignment constraints, which guarantee that each task is assigned to exactly one workstation. For example, constraint set 5.1.10 guarantees that task 1 can either be assigned to workstation 1 or 2 because $E_1 = 1$ and $L_1 = 2$, as shown in Table 5.1. Constraint sets 5.1.18 to 5.1.22 are the cycle time constraints, which ensure that the sum of task times assigned to each workstation does not exceed the theoretical cycle time—in this example, $C = 3.7$. Constraint sets 5.1.23 to 5.1.31 are the precedence relationship constraints. For example, constraint set 5.1.23 ensures that task 2 cannot be performed until task 1 is finished. Constraint sets 5.1.32 to 5.1.36 are the workstation constraints. For example, constraint set 5.1.32 makes sure that workstation 1 is used if there is task assigned to it. Constraint sets 5.1.37 to 5.1.46 guarantee that the earlier workstations are used if there is a task assigned to the workstation that proceeds it.

5.1.3 ORALBP: SAS Code for Assembly Line–Balancing Problem

ORALBP is a macro that solves the assembly line–balancing problem, the objective of which is to assign tasks to workstations while meeting the production requirements with the use of the minimum number of workstations (see program "sasor-5-1.sas"). The primary procedure used for the assembly line balancing–problem is PROC OPTMODEL. A full syntax of this procedure is available in Appendix 5.1.

Figure 5.2 illustrates the data flow in ORALBP. It shows:

- The workstation information matrix that is required for ORALBP, in which the tasks, task time, earliest and latest workstation, and succeeding tasks are specified.
- The macros (%data, %model, and %report).
- The results datasets that are available for print or can be used for further analysis.

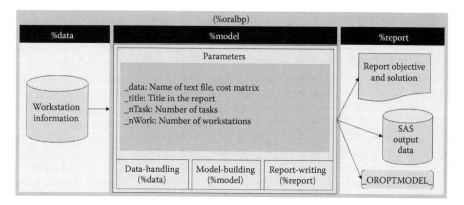

FIGURE 5.2
Data flow in ORALBP.

```
Data5_1.txt - Notepad                                    _ □ X
File  Edit  Format  View  Help
Task     time     Ei       Li        succ1    succ2    succ3
1        1        1        2         3        2        .
2        0.6      1        3         4        5        .
3        1.6      1        5         8        .        .
4        2.7      2        4         6        .        .
5        1.2      1        4         7        .        .
6        0.9      2        5         8        .        .
7        3.4      2        4         8        .        .
8        1.1      4        5         .        .        .
```

FIGURE 5.3
An example of a dataset, tasks and workstations information.

In the rest of this section, the procedure for solving an assembly line–balancing problem (ORALBP) in SAS, together with an example, is explained. ORALBP runs three macros: data-handling (%data), model-building (%model), and report-writing (%report).

5.1.4 ORALBP: Data-Handling Macro (%data)

This part of ORALBP processes the data into a format that is suitable for PROC OPTMODEL. The ORALBP requires one dataset containing the names of the tasks, durations, earliest and latest workstation, and succeeding tasks The dataset should be a .txt file, which is saved as "text tab delimited." The tasks' names must be given using numerical values. An example of data file is found in Figure 5.3.

One parameter needs to set before calling the data macro:

_data: Indicates the name and location of the data file (a text tab delimited file) and contains a workstation information matrix

```
* The data-handling macro;
%macro data;
* Import text tab delimited data file to the SAS data
file;
proc import
  datafile = &_data
  out = dataalb
  dbms = tab
  replace;
  getnames = yes;
 run;

data dataalb2 (drop = i);
set dataalb;
array W(&_nWork);
do i = 1 to &_nWork;
if Ei < = i < = Li then w(i) = 1; else w(i) = 0;
end;
run;
%mend data;
```

5.1.5 ORALBP: Model-Building Macro (%model)

This part of ORALBP calls PROC OPTMODEL to solve the model. Two parameters need to be set before calling the model macro:

_nTask: Defines the number of tasks

_nWork: Defines the maximum number of workstations

The SAS macro for model-building is as follows:

```
* The model-building macro;
%macro model;
* Starting OPTMODEL procedure;
proc optmodel;
* Define sets;
set TASKS = 1..&_nTask;
set WORKS = 1..&_nWork;

* Define variables;
var A{WORKS} binary > = 0;
var x{TASKS,WORKS} binary > = 0 ;
var z;
```

```
* Define parameters;
number Time{TASKS};
number EL{TASKS,WORKS};
number Succ1{TASKS};
number Succ2{TASKS};
number Succ3{TASKS};

* Load time (duration of each task);
read data dataalb2
into [_N_]
 Time[_N_] = col('time');

 * Load EL (Earliest/Latest workstations to which the
tasks can be assigned);
read data dataalb2
into [_N_]
 {i in WORKS} < EL[_N_,i] = col('W'||i) > ;

* Load Succ1 (First Immediate Successor);
read data dataalb2
into [_N_]
 Succ1[_N_] = col('Succ1');

* Load Succ2 (Second Immediate Successor);
read data dataalb2
into [_N_]
 Succ2[_N_] = col('Succ2');

* Load Succ3 (Third Immediate Successor);
read data dataalb2
into [_N_]
 Succ3[_N_] = col('Succ3');

* Define objective function ;
min obj = z;

* Define constrains;
con Objective: z = sum{i in WORKS} A[i];

con assignment_one_workstation {j in TASKS}: sum{i in
WORKS} EL[j,i]*x[j,i] = 1;

con cycle_time {i in WORKS}: sum{j in TASKS}
EL[j,i]*Time[j]*x[j,i] < = &_CycleT;

con precedence1 {j in TASKS: succ1[j] > 0}: sum{i in
WORKS} i*EL[j,i]*x[j,i]-sum{i in WORKS}
i*EL[succ1[j],i]*x[succ1[j],i] < = 0;
con precedence2 {j in TASKS: succ2[j] > 0}: sum{i in
WORKS} i*EL[j,i]*x[j,i]-sum{i in WORKS}
i*EL[succ2[j],i]*x[succ2[j],i] < = 0;
con precedence3 {j in TASKS: succ3[j] > 0}: sum{i in
WORKS} i*EL[j,i]*x[j,i]-sum{i in WORKS}
i*EL[succ3[j],i]*x[succ3[j],i] < = 0;
```

```
con workstation {i in WORKS}: sum{j in TASKS}
EL[j,i]*x[j,i]-sum{j in TASKS} EL[j,i]*A[i] < = 0;
con utilize_earlier {i1 in WORKS, i2 in WORKS: i2 > i1}:
A[i2]-A[i1] < = 0;

expand;
* Solve the model;
solve with MILP;
%put &_OROPTMODEL_;

* Create optimum values in a SAS dataset 'optimout';
create data optimout
from [TASKS WORKS]
 = {j in TASKS, i in WORKS}
amount = x;

*End of PROC OPTMODE;
quit;
%mend model;
```

5.1.6 ORALBP: Report-Writing Macro (%report)

The outputs from ORALBP include a report that contains all the information concerning the objective and results of the primal and dual solution of the ILP. The results are saved in the "optimout" dataset. The user can define appropriate names for each of this dataset before calling %oralbp macro
Another parameter that needs to be set before calling this macro is:

 _title: Gives a title in the output of the SAS

```
* The report-writing macro;
%macro report;
 * report the results in a tabulated form;
proc tabulate data = optimout;
title &_title;
class TASKS WORKS ;
var amount;
table TASKS = ."TASKS",
      WORKS*amount*sum
      / BOX = 'Assigning Task to Workstation' ;

run;
%mend report;
```

5.1.7 ORALBP: Macro (%oralbp)

To make the system as user friendly as possible, the %oralbp macro combines the data-handling, model-building, and report-writing codes.

```
* Assembly line balancing problem;
%macro oralbp;
  %data;
  %model;
  %report;
%mend oralbp;
```

In this code, the %oralbp macro is used to manage all the codes explained earlier, including data-handling, model-building, and report-writing. To come up with the results, the user needs to set up the parameters and run only one statement:

```
%ORALBP;
```

5.1.8 Instructions for Using ORALBP Macro

This section presents SAS code for the earlier example of the assembly line–balancing problem with eight tasks (see Table 5.1). The data are saved in file "data5_1.txt."

The user needs to set the parameters as required and run the following code:

```
* SAS macro for assembly line balancing problem;
%let _title = 'Example 5.1. A project network, Assembly
line balancing problem';
%let _data = 'c:/sasor/Data5_1.txt';
%let _nTask = 8;
%let _nWork = 5;
%oralbp;
```

This code determines the results based on the specified parameters and the cost matrix saved in the text file; it also also produces a macro variable (_ROPTMODEL_) at termination. The user can examine the results of this macro variable, whether PROC LP ran correctly, and what error or difficulty it encountered.

5.1.9 Sample Results from ORALBP Macro: Output from SAS

The results of running this code are presented in Figure 5.4. It shows that $x_{11} = x_{21} = x_{34} = x_{43} = x_{51} = x_{64} = x_{72} = x_{84} = 1$. This implies that tasks 1, 2, and 5 are assigned to workstation 1; task 7 is assigned to workstation 2; task 4 is assigned to workstation 3; and tasks 3, 6, and 8 are assigned to workstation 4. Therefore, the minimum number of workstation used, or the optimal solution value, is 4. The printout of the log file (Figure 5.5) shows that the

Assigning Task to Workstation	WORKS				
	1	2	3	4	5
	amount	amount	amount	amount	amount
	Sum	Sum	Sum	Sum	Sum
TASKS					
1	1.00	0.00	0.00	0.00	0.00
2	1.00	0.00	0.00	0.00	0.00
3	0.00	0.00	0.00	1.00	0.00
4	0.00	0.00	1.00	0.00	0.00
5	1.00	0.00	0.00	0.00	0.00
6	0.00	0.00	0.00	1.00	0.00
7	0.00	1.00	0.00	0.00	0.00
8	0.00	0.00	0.00	1.00	0.00

FIGURE 5.4
Result of %oralbp, solution summary.

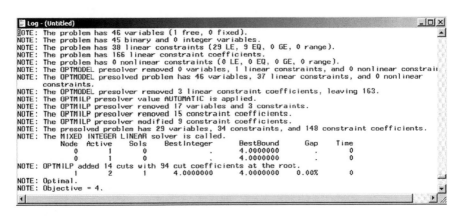

FIGURE 5.5
Log for %oralbp.

ORALBP procedure terminated successfully. The total computational time is 0.12 second.

5.1.10 Exercise

Use the codes developed in this chapter and solve the assembly line–balancing problem shown in Table 5.3.

It is assumed that the theoretical cycle time, C, is 1.8 units of time. The earliest workstation, E_i, and the latest workstation, L_i, to which task i can be assigned are listed in Table 5.4. Note that the maximum number of workstations available is four (i.e., $n = 4$).

Solution:

- Create the data in a text file (see "data5_1_exercise.txt").
- Run the following code:

```
* SAS macro for assembly line balancing problem: solution
to exercise 5.1.;
%let _title = 'A project network, Assembly line balancing
problem, solution to exercise 5.1.';
%let _data = 'c:/sasor/Data5_1_exercise.txt';
%let _nTask = 8;
%let _nWork = 4;
%let _CycleT = 1.8;
%oralbp;
```

TABLE 5.3

Precedence Relationships and
Time Requirements of a Work

Task	Immediate Predecessor(s)	Time
1	–	1.4
2	1	0.5
3	2	0.6
4	2	0.7
5	2	0.8
6	3	0.5
7	4, 5	1.0
8	6, 7	0.5

TABLE 5.4

Earliest and Latest
Workstations to Which Tasks
Can be Assigned

Task	E_i	L_i
1	1	1
2	2	2
3	2	4
4	2	3
5	2	3
6	2	4
7	3	4
8	4	4

The following solution is given by SAS:

Assigning Task to Workstation	WORKS			
	1	**2**	**3**	**4**
	amount	amount	amount	amount
	Sum	Sum	Sum	Sum
TASKS				
1	1.00	0.00	0.00	0.00
2	0.00	1.00	0.00	0.00
3	0.00	1.00	0.00	0.00
4	0.00	1.00	0.00	0.00
5	0.00	0.00	1.00	0.00
6	0.00	0.00	1.00	0.00
7	0.00	0.00	0.00	1.00
8	0.00	0.00	0.00	1.00

6

Traveling Salesman Problem

In this chapter, we present the traveling salesman problem and demonstrate how SAS/OR® can be applied to solve the problem to optimality.

6.1 Traveling Salesman Problem (TSP)

6.1.1 Concept of TSP

The *traveling salesman problem* (TSP) is one of the most widely studied integer programming problems and can be described as follows. Before visiting n distinct cities and then returning home, a salesman wants to determine the sequence of the travel so that the overall travel distance is minimized while visiting each city no more than once. Although the TSP is conceptually simple, it is difficult to obtain an optimal solution. In an n-city situation, any permutation of n cities yields a possible solution. As a consequence, $n!$ possible tours must be evaluated in the search space. By introducing decision variables x_{ij} to represent the tour of the salesman from city i to city j, the TSP can be formulated as shown in Model 6.1.1.

Model 6.1.1 Standard traveling salesman model

$$\text{Minimize } z = \sum_{i=1}^{n} \sum_{\substack{j=1 \\ j \neq i}}^{n} c_{ij} x_{ij} \tag{6.1.1}$$

subject to

$$\sum_{i=1}^{n} x_{ij} = 1 \quad j = 1, 2, \ldots, n; i \neq j \tag{6.1.2}$$

$$\sum_{j=1}^{n} x_{ij} = 1 \quad i = 1, 2, \ldots, n; i \neq j \tag{6.1.3}$$

$$u_i - u_j + n x_{ij} \leq n - 1 \quad i, j = 1, 2, \ldots, n; i \neq j \tag{6.1.4}$$

All $x_{ij} = 0$ or 1. All $u_i \geq 0$ and is a set of integers.

The distance between city i and city j is denoted as c_{ij}. Objective function 6.1.1 minimizes the total distance traveled in a tour. Constraint set 6.1.2 ensures that the salesman arrives at each city once. Constraint set 6.1.3 ensures that the salesman leaves each city once. Constraint set 6.1.4 is used to avoid a subtour. Because the decision variable x_{ij} is used, the solutions generated may form subtours, provided that constraint set 6.1.4 is not incorporated. Actually, the solutions consisting of subtours are infeasible because the travel sequence of the salesman is still unknown after all the decision variables x_{ij} are found. For example, the decision variables for a six-city problem are: $x_{12} = x_{23} = x_{31} = x_{46} = x_{65} = x_{54} = 1$. It is assumed that city 1 is the home city, which means that the salesman should start and end at this city. In this case, two subtours are formed:

- First subtour: city 1 → city 2 → city 3 → city 1
- Second subtour: city 4 → city 6 → city 5 → city 4

According to the first subtour, the salesman moves back to the home city, or city 1, after serving the customers in city 3. Which city is visited next after visiting all cities in the first subtour is not indicated. So this solution is unacceptable, and constraint set 6.1.4 in Model 6.1.1 must be included. Although constraint set 6.1.4 guarantees that the solution generated is feasible, it increases the complexity of the model because there are $n(n-1)$ constraints in this subtour elimination constraint.

6.1.2 Example of TSP

Table 6.1 shows a TSP that has six cities. It is assumed that city 1 is the home city. The upper-right corner of each cell in the tableau represents the travel distance, c_{ij}. Because it is not permitted to visit a city more than once, x_{ii} is discarded.

By introducing decision variables x_{ij} to represent the tour of the salesman from city i to city j, the TSP can be formulated as shown in Model 6.1.2.

Model 6.1.2 Example of formulation of TSP

$$\text{Minimize } 4\,x_{12} + 3\,x_{13} + 4\,x_{14} + 7\,x_{15} + 8\,x_{16}$$
$$+\, 4\,x_{21} + 4\,x_{23} + 3\,x_{24} + 7\,x_{25} + 6\,x_{26}$$
$$+\, 3\,x_{31} + 4\,x_{32} + 2\,x_{34} + 4\,x_{35} + 6\,x_{36}$$
$$+\, 4\,x_{41} + 3\,x_{41} + 2\,x_{43} + 4\,x_{45} + 4\,x_{46}$$
$$+\, 7\,x_{51} + 7\,x_{52} + 4\,x_{53} + 4\,x_{54} + 4\,x_{56}$$
$$+\, 8\,x_{61} + 6\,x_{62} + 6\,x_{63} + 4\,x_{64} + 4\,x_{65} \qquad (6.1.5)$$

subject to

$$x_{21} + x_{31} + x_{41} + x_{51} + x_{61} = 1 \qquad (6.1.6)$$

TABLE 6.1

A Traveling Salesman Tableau

City i	City j					
	1	**2**	**3**	**4**	**5**	**6**
1	– / –	4 / x_{12}	3 / x_{13}	4 / x_{14}	7 / x_{15}	8 / x_{16}
2	4 / x_{21}	– / –	4 / x_{23}	3 / x_{24}	7 / x_{25}	6 / x_{26}
3	3 / x_{31}	4 / x_{32}	– / –	2 / x_{34}	4 / x_{35}	6 / x_{36}
4	4 / x_{41}	3 / x_{42}	2 / x_{43}	– / –	4 / x_{45}	4 / x_{46}
5	7 / x_{51}	7 / x_{52}	4 / x_{53}	4 / x_{54}	– / –	4 / x_{56}
6	8 / x_{61}	6 / x_{62}	6 / x_{63}	4 / x_{64}	4 / x_{65}	– / –

$$x_{12} + x_{32} + x_{42} + x_{52} + x_{62} = 1 \tag{6.1.7}$$
$$x_{13} + x_{23} + x_{43} + x_{53} + x_{63} = 1 \tag{6.1.8}$$
$$x_{14} + x_{24} + x_{34} + x_{54} + x_{64} = 1 \tag{6.1.9}$$
$$x_{15} + x_{25} + x_{35} + x_{45} + x_{65} = 1 \tag{6.1.10}$$
$$x_{16} + x_{26} + x_{36} + x_{46} + x_{56} = 1 \tag{6.1.11}$$
$$x_{12} + x_{13} + x_{14} + x_{15} + x_{16} = 1 \tag{6.1.12}$$
$$x_{21} + x_{23} + x_{24} + x_{25} + x_{26} = 1 \tag{6.1.13}$$
$$x_{31} + x_{32} + x_{34} + x_{35} + x_{36} = 1 \tag{6.1.14}$$
$$x_{41} + x_{42} + x_{43} + x_{45} + x_{46} = 1 \tag{6.1.15}$$
$$x_{51} + x_{52} + x_{53} + x_{54} + x_{56} = 1 \tag{6.1.16}$$
$$x_{61} + x_{62} + x_{63} + x_{64} + x_{65} = 1 \tag{6.1.17}$$
$$u_1 - u_2 + 6 x_{12} \leq 5 \tag{6.1.18}$$
$$u_1 - u_3 + 6 x_{13} \leq 5 \tag{6.1.19}$$
$$u_1 - u_4 + 6 x_{14} \leq 5 \tag{6.1.20}$$
$$u_1 - u_5 + 6 x_{15} \leq 5 \tag{6.1.21}$$
$$u_1 - u_6 + 6 x_{16} \leq 5 \tag{6.1.22}$$
$$u_2 - u_1 + 6 x_{21} \leq 5 \tag{6.1.23}$$
$$u_2 - u_3 + 6 x_{23} \leq 5 \tag{6.1.24}$$

$$u_2 - u_4 + 6 x_{24} \leq 5 \tag{6.1.25}$$

$$u_2 - u_5 + 6 x_{25} \leq 5 \tag{6.1.26}$$

$$u_2 - u_6 + 6 x_{26} \leq 5 \tag{6.1.27}$$

$$u_3 - u_1 + 6 x_{31} \leq 5 \tag{6.1.28}$$

$$u_3 - u_2 + 6 x_{32} \leq 5 \tag{6.1.29}$$

$$u_3 - u_4 + 6 x_{34} \leq 5 \tag{6.1.30}$$

$$u_3 - u_5 + 6 x_{35} \leq 5 \tag{6.1.31}$$

$$u_3 - u_6 + 6 x_{36} \leq 5 \tag{6.1.32}$$

$$u_4 - u_1 + 6 x_{41} \leq 5 \tag{6.1.33}$$

$$u_4 - u_2 + 6 x_{42} \leq 5 \tag{6.1.34}$$

$$u_4 - u_3 + 6 x_{43} \leq 5 \tag{6.1.35}$$

$$u_4 - u_5 + 6 x_{45} \leq 5 \tag{6.1.36}$$

$$u_4 - u_6 + 6 x_{46} \leq 5 \tag{6.1.37}$$

$$u_5 - u_1 + 6 x_{51} \leq 5 \tag{6.1.38}$$

$$u_5 - u_2 + 6 x_{52} \leq 5 \tag{6.1.39}$$

$$u_5 - u_3 + 6 x_{53} \leq 5 \tag{6.1.40}$$

$$u_5 - u_4 + 6 x_{54} \leq 5 \tag{6.1.41}$$

$$u_5 - u_6 + 6 x_{56} \leq 5 \tag{6.1.42}$$

$$u_6 - u_1 + 6 x_{61} \leq 5 \tag{6.1.43}$$

$$u_6 - u_2 + 6 x_{62} \leq 5 \tag{6.1.44}$$

$$u_6 - u_3 + 6 x_{63} \leq 5 \tag{6.1.45}$$

$$u_6 - u_4 + 6 x_{64} \leq 5 \tag{6.1.46}$$

$$u_6 - u_5 + 6 x_{65} \leq 5 \tag{6.1.47}$$

All $x_{ij} = 0$ or 1. All $u_i \geq 0$ and is a set of integers.

Objective function 6.1.5 calculates the total travel distance of the salesman for serving all customers in the cities. If the moving speed is incorporated, then the objective can be the minimization of the total travel time. Constraint sets 6.1.6 to 6.1.11 guarantee that exactly one city must be visited immediately before city j. Constraint sets 6.1.12 to 6.1.17 ensure that exactly one city must be visited immediately after city i. Constraint sets 6.1.18 to 6.1.47 are known as the *subtour elimination constraints*. Because the model contains binary variables (i.e., x_{ij}) and integer variables (i.e., u_i), Model 6.1.2 can be regarded as a *pure integer linear programming model*.

TABLE 6.2

Definitions of the TSP Variants

TSP Variants	Definitions
Asymmetric TSP	The bidirectional distances between a pair of customers are not necessarily identical.
Generalized TSP	The customers are divided into several groups, and not all customers need to be visited.
Multiple TSP	More than one salesman is allowed to be used, and each customer is served by a salesman only.
Period TSP	The vehicle has to visit each customer several times over a given period of time.
Pickup-and-delivery TSP	The vehicle may both receive products from the customers and deliver products to the customers at the same time.
Probabilistic TSP	Each customer has a predetermined probability of requiring a visit.
TSP with precedence constraints	There exist a delivering order between the customers.
TSP with time window	The vehicles must arrive at the customers before the latest arrival time, but arriving before the earliest arrival time results in waiting.

In many real-life situations, a TSP is not solved optimally by exact algorithms, especially when the problem size is huge. Instead, it is solved by heuristics, such as the nearest neighbor heuristic. The principle of the nearest neighbor heuristic is to start with the first city randomly, then to select the next city as close as possible to the previous one from those unselected cities to form the travel sequence until all cities are selected.

The most common TSP is known as the Euclidean TSP, in which distance matrix c is expected to be symmetrical, (i.e., $c_{ij} = c_{ji}$ for all i, j),and to satisfy the triangle inequality (i.e., $c_{ik} \leq c_{ij} + c_{jk}$ for all distinct i, j, k). There are extensive variations of the TSP, including the asymmetric TSP, generalized TSP, multiple TSP, period TSP, pickup-and-delivery TSP, probabilistic TSP, TSP with precedence constraints, and TSP with time window. The definitions of all TSP variants are described in Table 6.2.

6.1.3 ORTSP: SAS Code for TSP

ORTSP is a macro that solves TSPs in which a salesman wants to visit n distinct cities and then return home (see program "sasor_6_1.sas"). The aim is to determine the sequence of the travel so that the overall travel distance is minimized while visiting each city not more than once. The primary procedure used for a TSP is PROC GA.

Figure 6.1 illustrates the data flow in the ORTSP. It shows:

- The distance matrix that is required for ORTSP, in which the distance between each pair of traveling points is specified

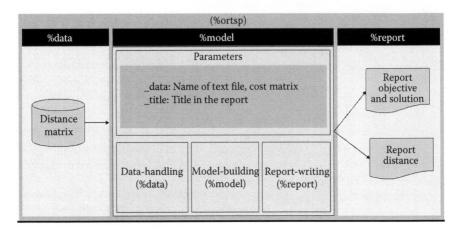

FIGURE 6.1
Data flow in ORTSP.

```
Data6_1.txt - Notepad                                    _|□|X|
File   Edit   Format   View   Help
city      city1     city2     city3     city4     city5     city6
city1     0         4         3         4         7         8
city2     4         0         4         3         7         6
city3     3         4         0         2         4         6
city4     4         3         2         0         4         4
city5     7         7         4         4         0         4
city6     8         6         6         4         4         0
```

FIGURE 6.2
An example of data set, distance matrix.

- The macros (%data, %model, and %report)
- The results data sets that are available for print or can be used for further analysis

In the rest of this section, the procedure for solving a TSP (ORTSP) in SAS, together with an example, is explained. ORTSP runs three macros: data-handling (%data), model-building (%model), and report-writing (%report).

6.1.4 ORTSP: Data-Handling Macro (%data)

This part of ORTSP processes the data into a format that is suitable for PROC GA. The ORTSP requires one dataset containing the names of traveling points in the first row, and distance matrix. The dataset should be a .txt file, which is saved as "text tab delimited." The names of traveling points must start with a letter and may contain up to 50 characters. The name of traveling points should be listed in the first column. An example of data file is shown in Figure 6.2.

One parameter needs to be set before calling the data macro:

_data: Indicates the name and location of the data file (a text tab delimited file) and contains the distance matrix

```
* The data-handling macro;
%macro data;
* Import text tab delimited data file to SAS data file;
  proc import
        datafile = &_data
        out = dtsp1
        dbms = tab
        replace;
        getnames = yes;
  run;

  data dtsp (drop = &_travelP);
    set dtsp1;
  run;
%mend data;
```

6.1.5 ORTSP: Model-Building Macro (%model)

This part of ORTSP calls PROC GA to solve the model. Two parameters need to be set before calling the model macro:

_tail: Defines a variable in the data file for origin

_travelP: Defines the name of the variable to identify traveling points

The SAS macro for model-building is as follows:

```
* The model-building macro;
%macro model;
  proc ga Matrix1 = dtsp;
  call SetEncoding ('S6');
  call SetObj ('TSP',0,'distances',Matrix1);
  call SetCross ('Order');
  call Initialize ('DEFAULT', 6);
  call ContinueFor (85);
  run;
%mend model;
```

6.1.6 ORTSP: Report-Writing Macro (%report)

The outputs from ORTSP include a report containing the objective and the solution. One parameter needs to be set before calling this %report macro:

_title: Gives a title in the output of the SAS

```
* The report-writing macro;
%macro report;
 title &_title;
 proc print;
 run;
%mend report;
```

6.1.7 ORTSP: Macro (%ortsp)

To make the system as user friendly as possible, the %ortsp macro combines the data-handling, model-building, and report-writing codes.

```
* An SAS macro for a traveling salesman problem;
%macro ortsp;
 %data;
 %model;
 %report;
%mend ortsp;
```

In this code, the %ortsp macro is used to manage all the codes explained earlier, including data-handling, model-building, and report-writing. To get the results, user needs to set up the parameters and run only one statement:

```
%ORTSP;
```

6.1.8 Instructions for Using ORTSP Macro

This section presents SAS code for the earlier example of the TSP with six cities as shown in Table 6.1. The data are saved in file "data6_1.txt." The user needs to set the parameters as required and run the following code:

```
* SAS procedure for traveling salesman problem;
%let _data = 'c:/sasor/data6_1.txt';
%let _travelP = city;
%let _title = 'Example 6.1: Traveling salesman problem';
%ortsp;
```

This code determines the results based on the specified distances.

6.1.9 Sample Results from ORTSP Macro: Output from SAS

The results of running this code are presented in Figure 6.3, and the example data file is shown in Figure 6.4. In Figure 6.3, the column "Element" refers to the travel sequence, whereas the column "Value" refers to the city. Therefore,

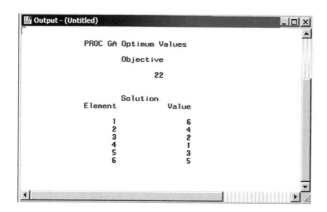

FIGURE 6.3
Result of %ORTSP, objective and results.

Obs	city1	city2	city3	city4	city5	city6
1	0	4	3	4	7	8
2	4	0	4	3	7	6
3	3	4	0	2	4	6
4	4	3	2	0	4	4
5	7	7	4	4	0	4
6	8	6	6	4	4	0

FIGURE 6.4
Result of % ORTSP, data file.

the first city to be visited is city 6. The entire travel sequence of the salesman will be city 6 → city 4 → city 2 → city 1 → city 3 → city 5 → city 6. The optimal solution value is 22 units. Note that the code we use in this section is GA, which is a heuristic algorithm. Therefore, it cannot guarantee that the solution is globally optimal.

6.1.10 Exercise

Use the codes developed in this chapter and solve the following TSP as shown in Table 6.3.
    Solution:

- Create the data in a text file (see "data6_1_exercise.txt")
- Run the following code (see program "sasor_6_1_exercise.sas"):

```
* SAS macro for traveling salesman problem: solution to
exercise 6.1.;
%let _title = 'Traveling salesman problem: solution to
exercise 6.1.';%let _data = 'c:/sasor/data6_1_exercise.
txt';%let _travelP = city;%let _title = ' Traveling salesman
problem, solution toexercise';%ortsp;
```

TABLE 6.3

A Traveling Salesman Exercise

City i	City j					
	1	**2**	**3**	**4**	**5**	**6**
1	– –	5 x_{12}	7 x_{13}	9 x_{14}	8 x_{15}	6 x_{16}
2	5 x_{21}	– –	7 x_{23}	10 x_{24}	9 x_{25}	9 x_{26}
3	7 x_{31}	7 x_{32}	– –	8 x_{34}	6 x_{35}	10 x_{36}
4	9 x_{41}	10 x_{42}	8 x_{43}	– –	10 x_{45}	7 x_{46}
5	8 x_{51}	9 x_{52}	6 x_{53}	10 x_{54}	– –	8 x_{56}
6	6 x_{61}	9 x_{62}	10 x_{63}	7 x_{64}	8 x_{65}	–

The following solution is given by SAS:

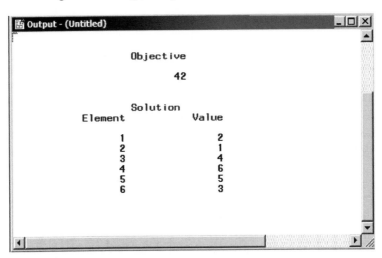

7

Printed Circuit Board Production Planning

In this chapter, we present a wide range of printed circuit board production-planning problems and demonstrate how SAS/OR® can be applied to solve the problems to optimality. The problems include PCB assembly line assignment, PCB component allocation, and PCB component sequencing for pick-and-place machines.

7.1 Printed Circuit Board (PCB) Assembly Line Assignment Problem

7.1.1 Concept of PCB Assembly Line Assignment Problem

Printed circuit boards (PCBs) are extensively used in a variety of products today. These products include computers, mobile phones, and audiovisual equipment. A PCB consists of a pattern of electrical traces etched from copper that is laminated on an insulated base that is typically rigid fiberglass. The PCB serves as the interconnection device for electrical currents traveling on the board and the discrete electronic components that are essential to the functioning of an electronic product. Components—from a few hundred to some thousands—can be assembled on a single PCB (Ho and Ji, 2006b).

The process of assembling electronic components on a PCB is called *PCB assembly*. It can be classified into two categories: plated-through-hole (PTH) technology and surface mount technology (SMT). PTH technology is used with products for which board size is not a major concern. The components are inserted into the holes drilled through the PCB. Then the connections are soldered on the underside of the PCB between the component lead and the PCB pad. However, requirements from consumers, such as smaller product size and greater functionality and reliability, have lead to SMT replacing PTH technology. The configuration and the size of surface mount components have permitted mounting a large number of components on a single PCB (Ho and Ji 2006b).

A PCB manufacturing company receives production orders for many distinct products every month and may have several SMT assembly lines. The production volume for each type of product is high. The scheduler has to determine which product should be produced on which line and what quantity of the product should be produced on which line so that the production cost is minimized. A product or board type may be produced on one line only or on more than one line, depending on the product order and the availability of the assembly lines. Figure 7.1 shows n board types to be assigned to three assembly lines (Ho and Ji 2006b).

There are two types of placement machines in the SMT assembly line, as shown in Figure 7.1. Each type of machine possesses its own peculiarities as well as the operation. The first type of machine is called the *sequential pick-and-place (PAP) machine*. In this machine, components of the same type are stored in a single stationary feeder and the PCB is secured on a fixed working table. During the placement operation, the assembly head travels to pick up one component at a time from a feeder and then places it on the stationary board. The PAP machine can achieve high accuracy. Moreover, it is suitable for operations with large components, such as integrated circuits (Ho and Ji 2004; 2005).

The second type of machine is the *concurrent chip shooter (CS) machine*. It possesses an X-Y table carrying a PCB, a feeder carrier with several feeders holding components, and a rotary turret with multiple assembly heads to pick up and place components. Each assembly head has several nozzles of different sizes. A large nozzle is used to pick up and place large components. The major advantage of the CS machine is its high speed because the pickup and placement operations are performed concurrently. However, it is only preferable for operations with small components, such as chip resistors. Because the placement of smaller components is given priority, this type of machine is arranged before the PAP machine in the assembly line (Ho and Ji 2003; 2006a).

According to Figure 7.1, the line configurations are different from each other. If these three lines are assigned to produce a board type, the times

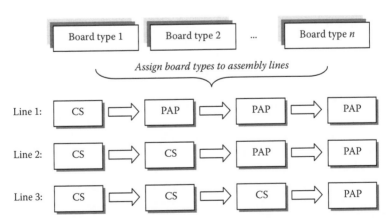

FIGURE 7.1
An example of a PCB assembly line assignment problem. CS, chip shooter; PAP, pick and place.

required by these three lines to produce 1 unit of the board type will not be identical, nor will the efficiency and the cost.

The assignment of board types to multiple SMT assembly lines is addressed as a line assignment problem. Here, the problem is formulated as the generalized transportation problem (GTP). Actually, the GTP is an extension of the linear transportation problem, discussed in Section 2.1. However, it does not have the integrality property, which means the linear programming solution (like the simplex method) cannot guarantee to be integer. The GTP (Balas and Ivanescu 1964; Lourie, 1964) or the machine-loading problem (Eisemann 1964), proposed by Ferguson and Dantzig (1956), has been studied for a long time because of its wide applicability (Eisemann 1964; Ji et al. 1994).

The mathematical model of the GTP assumes that m assembly lines are available for the production of n board types for which the production volume is high. Here, production means that each board is processed in a single line instead of by a specified sequence of lines. Furthermore, a board can be produced in any line. When line i is assigned to produce board type j, it requires a_{ij} (> 0) hours and costs c_{ij} (> 0) dollars to make 1 unit j. In addition, line i has a maximum of s_i hours available for production while board type j has a volume requirement of d_j. The problem here is to determine the quantity x_{ij} of board type j to be produced on line i so as to minimize the total production cost. A pure integer linear programming model can be formulated as shown in Model 7.1.1.

Model 7.1.1 Standard PCB assembly line assignment model

$$\text{Minimize } z = \sum_{i=1}^{m} \sum_{j=1}^{n} c_{ij} x_{ij} \tag{7.1.1}$$

subject to

$$\sum_{j=1}^{n} a_{ij} x_{ij} \leq S_i \qquad i = 1, 2, \ldots, m \tag{7.1.2}$$

$$\sum_{i=1}^{m} x_{ij} = d_j \qquad j = 1, 2, \ldots, n \tag{7.1.3}$$

$x_{ij} \geq 0$ and is a set of integers.

Objective function 7.1.1 minimizes the total production cost. Constraint set 7.1.2 is necessary due to the limited available time, which means that each line must be operated within the fixed time period. Constraint set 7.1.3 arises from the product volume requirement. Model 7.1.1 is referred to as the GTP.

7.1.2 Example of PCB Assembly Line Assignment Problem

Table 7.1 shows a line assignment problem tableau, which has four lines and five types of boards (Ho and Ji 2006b). The upper-left corner in cell (i, j) indicates the unit time, a_{ij}, and the upper-right corner represents the unit cost, c_{ij}.

By introducing decision variables x_{ij} to represent the amount of board type j to be produced on line i, the PCB line assignment problem in Table 7.1 can be formulated as shown in Model 7.2.1.

Model 7.1.2 Example of formulation of PCB assembly line assignment problem

$$\text{Minimize} \quad 3\, x_{11} + 6\, x_{12} + 6\, x_{13} + 7\, x_{14} + 4\, x_{15}$$

$$+ 6\, x_{21} + 5\, x_{22} + 6\, x_{23} + 15\, x_{24} + 8\, x_{25}$$

$$+ 5\, x_{31} + 3\, x_{32} + 10\, x_{33} + 6\, x_{34} + 7\, x_{35}$$

$$+ 4\, x_{41} + 4\, x_{42} + 7\, x_{43} + 8\, x_{44} + 9\, x_{45} \qquad (7.1.4)$$

subject to

$$5\, x_{11} + 2\, x_{12} + 2\, x_{13} + 5\, x_{14} + 6\, x_{15} \leq 48000 \qquad (7.1.5)$$

$$2\, x_{21} + 7\, x_{22} + 8\, x_{23} + 7\, x_{24} + 6\, x_{25} \leq 48000 \qquad (7.1.6)$$

$$4\, x_{31} + 2\, x_{32} + 2\, x_{33} + x_{34} + 6\, x_{35} \leq 48000 \qquad (7.1.7)$$

$$3\, x_{41} + 2\, x_{42} + 4\, x_{43} + 5\, x_{44} + 2\, x_{45} \leq 48000 \qquad (7.1.8)$$

$$x_{11} + x_{21} + x_{31} + x_{41} = 12000 \qquad (7.1.9)$$

TABLE 7.1

A PCB Line Assignment Problem Tableau

Line i	Board type j										s_i
	1		**2**		**3**		**4**		**5**		
1	5 x_{11}	3	2 x_{12}	6	2 x_{13}	6	5 x_{14}	7	6 x_{15}	4	48000
2	2 x_{21}	6	7 x_{22}	5	8 x_{23}	6	7 x_{24}	15	6 x_{25}	8	48000
3	4 x_{31}	5	2 x_{32}	3	2 x_{33}	10	1 x_{34}	6	6 x_{35}	7	48000
4	3 x_{41}	4	2 x_{42}	4	4 x_{43}	7	5 x_{44}	8	2 x_{45}	9	48000
d_j	12000		8000		10000		8000		6000		

$$x_{12} + x_{22} + x_{32} + x_{42} = 8000 \qquad (7.1.10)$$

$$x_{13} + x_{23} + x_{33} + x_{43} = 10000 \qquad (7.1.11)$$

$$x_{14} + x_{24} + x_{34} + x_{44} = 8000 \qquad (7.1.12)$$

$$x_{15} + x_{25} + x_{35} + x_{45} = 6000 \qquad (7.1.13)$$

$$x_{ij} \geq 0 \text{ and is a set of integers}$$

Constraint sets 7.1.5 to 7.1.8 are the availability constraints. There is one such constraint for each assembly line. For example, constraint set 7.1.5 ensures that the total production time in line 1 must not exceed the limited available time—that is, 48,000 units of time. Constraint sets 7.1.9 to 7.1.13 are the requirement constraints. Similarly, there is one such constraint for each board type. For example, constraint set 7.1.9 ensures that the total amount of board type 1 must be assigned to the assembly line or lines.

7.1.3 ORALA: SAS Code for PCB Assembly Line Assignment Problem

ORALA is a macro that solves PCB assembly line assignment problems, the objective of which is to minimize the total production cost (see program "sasor-7-1.sas"). The primary procedure used for the PCB assembly line assignment problem is PROC OPTMODEL.

Figure 7.2 illustrates the data flow in the ORALA. It shows:

- The time matrix and the cost matrix that are required for ORALA, in which the unit time and the unit cost of any line *i* and board *j* are specified

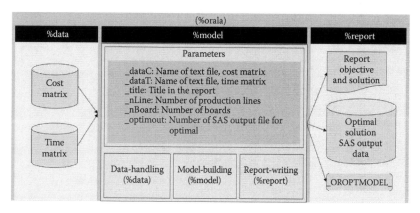

FIGURE 7.2
Data flow in ORALA.

- The macros (%data, %model, and %report)
- The results datasets that are available for print or can be used for further analysis

In the rest of this section, the procedure for solving the PCB assembly line assignment problem (ORALA) in SAS, together with an example, is explained. ORALA runs three macros: data-handling (%data), model-building (%model), and report-writing (%report).

7.1.4 ORALA: Data-Handling Macro (%data)

This part of ORALA processes the data into a format that is suitable for PROC OPTMODEL. ORALA requires two data sets containing the names of the lines and boards and the cost matrix and the time matrix. The data set should be a .txt file, which is saved as "text tab delimited." The names of the lines and boards must start with a letter and may contain up to 50 characters. The boards' names must be listed in the first row of the data file. The lines' name should be listed in the first column. Number of lines and number of boards should be given to the macro before calling it. Example data files are seen in Figures 7.3 and 7.4.

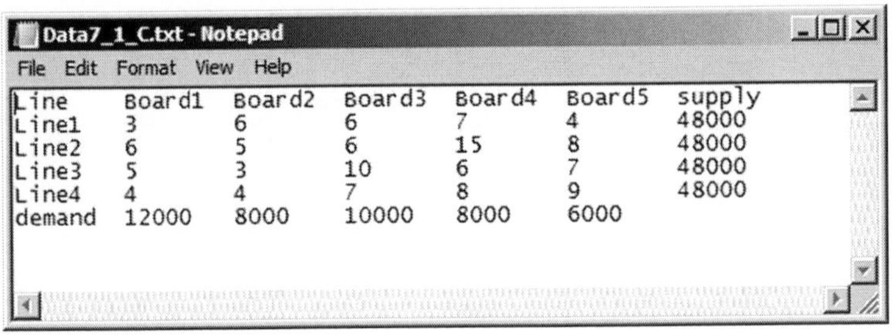
Data7_1_C.txt - Notepad
File Edit Format View Help

Line	Board1	Board2	Board3	Board4	Board5	supply
Line1	3	6	6	7	4	48000
Line2	6	5	6	15	8	48000
Line3	5	3	10	6	7	48000
Line4	4	4	7	8	9	48000
demand	12000	8000	10000	8000	6000	

FIGURE 7.3
An example of a dataset, cost matrix.

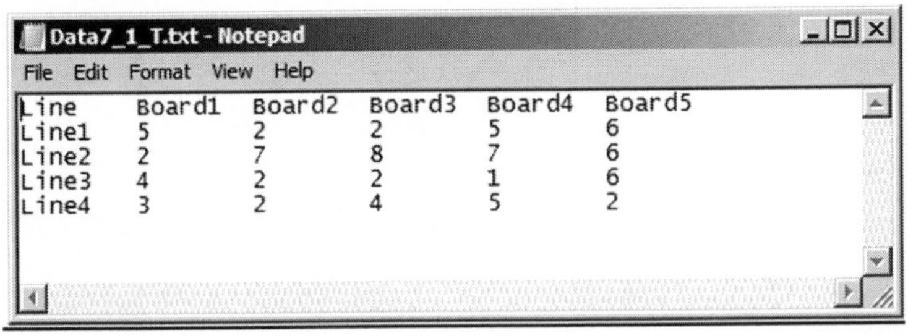
Data7_1_T.txt - Notepad
File Edit Format View Help

Line	Board1	Board2	Board3	Board4	Board5
Line1	5	2	2	5	6
Line2	2	7	8	7	6
Line3	4	2	2	1	6
Line4	3	2	4	5	2

FIGURE 7.4
An example of dataset, time matrix.

Two parameters need to be set before calling the data macro:

_dataC: Indicates the name and location of the cost data file (a text tab delimited file) and contains cost matrix

_dataT: Indicates the name and location of the time data file (a text tab delimited file) and contains time matrix

```
* The data-handling macro;
%macro data;
* Import text tab delimited data file to SAS data file;
proc import
        datafile = &_dataC
        out = dataC
        dbms = tab
        replace;
        getnames = yes;
  run;
proc import
        datafile = &_dataT
        out = dataT
        dbms = tab
        replace;
        getnames = yes;
  run;
%mend data;
```

7.1.5 ORALA: Model-Building Macro (%model)

This part of ORALA calls PROC OPTMODEL to solve the model. Two parameters need to be set before calling the model macro:

_nLine: Identifies the number of lines

_nBoard: Identifies the number of boards

The SAS macro for model-building is as follows:

```
* The model-building macro;
%macro model;

* Starting OPTMODEL Procedure;
proc optmodel;

* Define sets;
set LINES = 1..&_nLine;
set BOARDS = 1..&_nBoard;
```

```
* Define variables;
var X{LINES, BOARDS} integer > = 0;

* Define parameters;
number cost{LINES, BOARDS};
number time{LINES, BOARDS};
number s{LINES};
number d{BOARDS};

* Load the time matrix;
read data dataT
into [_N_]
{j in BOARDS} < time[_N_,j] = col('board'||j) > ;

* Load the cost matrix;
read data dataC (where = (line ne "demand"))
into [_N_]
{j in BOARDS} < cost[_N_,j] = col('board'||j) > ;

* Load the demand array;
read data dataC (where = (line eq "demand"))
into
{j in BOARDS} < d[j] = col('board'||j) > ;

* Load the supply array;
read data dataC (where = (line ne "demand"))
into [_N_] s[_N_] = col("supply");

* Define objective function;
min obj = sum{i in LINES, j in BOARDS} cost[i,j]*X[i,j];

* Define constraints;
con supply_line{i in LINES}:
 sum{j in BOARDS} time[i,j]*X[i,j] < = s[i];

con demand_board{j in BOARDS}:
 sum{i in LINES} X[i,j] = d[j];

* Solve the model;
solve with milp;
%put &_OROPTMODEL_;

* Expand the model;
expand;

* Create optimum values in a SAS dataset 'optimout';
create data &_optimout
from [LINES BOARDS]
 = {i in LINES, j in BOARDS: x[i,j]^ = 0}
amount = x ;
```

```
* End of OPTMODEL Procedure;
quit;
%mend model;
```

7.1.6 ORALA: Report-Writing Macro (%report)

The outputs from ORALA include one report in the form of a table of boards and lines. This information is also saved in the "optimout" dataset. The user can define the appropriate names for the datasets before calling %orala macro:

_optimout: Identifies the name of the SAS output file for optimal solution

Another parameter needs to be set before calling this macro:

_title: Gives a title in the output of the SAS

```
* The report-writing macro;
%macro report;
* Report the results in a tabulated form;
proc tabulate data = &_optimout;
title &_title;
class LINES BOARDS ;
var amount;
table LINES = " Line",
BOARDS*amount*sum
/ BOX = 'Assigning of Boards to Lines';
run;
%mend report;
```

7.1.7 ORALA: Macro (%orala)

To make the system as user friendly as possible, the %orala macro combines the data-handling, model-building, and report-writing codes.

```
* A SAS macro for a line assignment problem;
%macro orala;
 %data;
 %model;
 %report;
%mend orala;
```

In this code, the %orala macro is used to manage all the codes explained earlier, including data-handling, model-building, and report-writing. To get the results, the user needs to set up the parameters and run only one statement:

```
%orala;
```

7.1.8 Instructions for Using ORALA Macro

This section presents SAS code for the earlier example of the PCB assembly line assignment problem with four lines and five boards as shown in Table 7.1. The data are saved in the files "data7_1_C.txt" and "data7_1_T.txt." The user needs to set the parameters as required and run the following code:

```
* Using PROC OPTMODEL for Printed Circuit Board (PCB)
Assembly Line Assignment Problem;
option nodate ;
option nonumber ;
  %let _title = 'Example 7.1: A line assignment problem
using PROC OPTMODEL';
  %let _dataC = 'c:/sasor/Data7_1_C.txt';
  %let _dataT = 'c:/sasor/Data7_1_T.txt';
  %let _nLine = 4;
  %let _nBoard = 5;
  %let _optimout = mysolution;
```

This code determines the results based on the specified parameters and the cost and time matrices saved in the text files; it also produces a macro variable (_OROPTMODEL_) at termination. The user can examine the result of this macro variable, examine whether PROC OPTMODEL ran correctly, and examine what error or difficulty it encountered. A summary of information, including the objective value at optimum level and the status of _OROPTMODEL_, can be seen in the log file as shown in Figure 7.5.

7.1.9 Sample Results from ORALA Macro: Output from SAS

The results of running this code are presented in Figures 7.5 and 7.6, which show the results of the primal model and the dual model, respectively. In Figure 7.5, there are 20 decision variables, starting from Boardall1 to Boardall20. Boardall1 is equivalent to x_{11} in Model 7.1.2, whereas Boardall6 is equivalent to x_{21}. The optimal solution is $x_{11} = 800$, $x_{13} = 4000$, $x_{15} = 6000$, $x_{23} = 6000$, $x_{32} = 8000$, $x_{34} = 8000$, and $x_{41} = 11200$. The optimal solution value is $203,200. Figure 7.7 shows the output from the SAS log file.

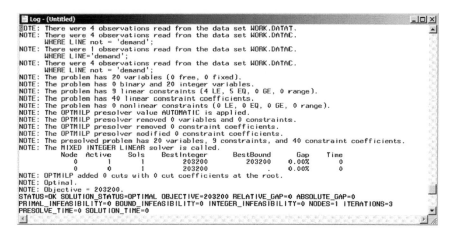

FIGURE 7.5
Results of %ORALA, solution of dual program.

Assigning of Boards to Lines	BOARDS				
	1	2	3	4	5
	amount	amount	amount	amount	amount
	Sum	Sum	Sum	Sum	Sum
Line					
1	800.00	.	4000.00	.	6000.00
2	.	.	6000.00	.	.
3	.	8000.00	.	8000.00	.
4	11200.00	.	.	.	.

FIGURE 7.6
Results of %ORALA

Note that the mixed integer linear programming (MILP) solver is used in this macro. Because we used the "expand" statement in the %model macro, the model is printed in the output window, as shown in Figure 7.7.

7.1.10 Exercise

Use the codes developed in this chapter and solve the assembly line assignment problem in Table 7.2.
Solution:

- Create the data in text files (see "data7_1_C_exercise.txt" and "data7_1_T_exercise.txt").
- Run the following code:

```
Var X[1,1] INTEGER >= 0
Var X[1,2] INTEGER >= 0
Var X[1,3] INTEGER >= 0
Var X[1,4] INTEGER >= 0
Var X[1,5] INTEGER >= 0
Var X[2,1] INTEGER >= 0
Var X[2,2] INTEGER >= 0
Var X[2,3] INTEGER >= 0
Var X[2,4] INTEGER >= 0
Var X[2,5] INTEGER >= 0
Var X[3,1] INTEGER >= 0
Var X[3,2] INTEGER >= 0
Var X[3,3] INTEGER >= 0
Var X[3,4] INTEGER >= 0
Var X[3,5] INTEGER >= 0
Var X[4,l] INTEGER >= 0
Var X[4,2] INTEGER >= 0
Var X[4,3] INTEGER >= 0
Var X[4,4] INTEGER >= 0
Var X[4,5] INTEGER >= 0
Minimize obj=3*X[l,1] + 6*X[1,2] + 6*X[1,3] + 7*X[1,4] + 4*X[1,5] + 6*X[2,1] + 5*X[2,2] +
6*X[2,3] + 15*X[2,4] + 8*X[2,5] + 5*X[3,1] + 3*X[3,2] + 10*X[3,3] + 6*X[3,4] + 7*X[3,5] +
4*X[4,1] + 4*X[4,2] + 7*X[4,3] + 8*X[4,4] + 9*X[4,5]
Constraint supply_line[1]: 5*X[1,1] + 2*X[1,2] + 2*X[1,3] + 5*X[1,4] + 6*X[1,5] <= 48000
Constraint supply_line[2]: 2*X[2,1] + 7*X[2,2] + 8*X[2,3] + 7*X[2,4] + 6*X[2,5] <= 48000
Constraint supply_line[3]: 4*X[3,1] + 2*X[3,2] + 2*X[3,3] + X[3,4] + 6*X[3,5] <= 48000
Constraint supply_line[4]: 3*X[4,1] + 2*X[4,2] + 4*X[4,3] + 5*X[4,4] + 2*X[4,5] <= 48000
Constraint demand_board[1]: X[1,1] + X[2,1] + X[3,1] + X[4,1] = 12000
Constraint demand_board[2]: X[1,2] + X[2,2] + X[3,2] + X[4,2] = 8000
Constraint demand_board[3]: X[1,3] + X[2,3] + X[3,3] + X[4,3] = 10000
Constraint demand_board[4]: X[1,4] + X[2,4] + X[3,4] + X[4,4] = 8000
Constraint demand_board[5]: X[1,5] + X[2,5] + X[3,5] + X[4,5] = 6000
```

FIGURE 7.7
The model used in this section as reported in the Output Window

```
* SAS macro for line assignment problem: solution to
exercise 7.1.;
%let _title = 'A line assignment problem using PROC
OPTMODEL: solution to exercise 7.1.';
 %let _dataC = 'c:/sasor/Data7_1_C_exercise.txt';
 %let _dataT = 'c:/sasor/Data7_1_T_exercise.txt';
 %let _nLine = 3;
 %let _nBoard = 5;
 %let _optimout = mysolution;
```

%orala;

TABLE 7.2

A PCB Line Assignment Exercise

Line i	Board type j					s_i
	1	2	3	4	5	
1	6 x_{11} 4	1 x_{12} 7	3 x_{13} 5	7 x_{14} 6	9 x_{15} 8	48000
2	4 x_{21} 9	6 x_{22} 6	7 x_{23} 7	4 x_{24} 10	3 x_{25} 8	48000
3	2 x_{31} 6	3 x_{32} 8	6 x_{33} 10	1 x_{34} 9	2 x_{35} 7	48000
d_j	3000	6000	9000	6000	3000	

The following solution is given by SAS:

Assigning of Boards to Lines	BOARDS				
	1	2	3	4	5
	amount	amount	amount	amount	amount
	Sum	Sum	Sum	Sum	Sum
Line					
1	.	.	9000.00	3000.00	.
2	.	6000.00	.	.	.
3	3000.00	.	.	3000.00	3000.00

```
Log - (Untitled)
NOTE: The problem has 15 variables (0 free, 0 fixed).
NOTE: The problem has 0 binary and 15 integer variables.
NOTE: The problem has 8 linear constraints (3 LE, 5 EQ, 0 GE, 0 range).
NOTE: The problem has 30 linear constraint coefficients.
NOTE: The problem has 0 nonlinear constraints (0 LE, 0 EQ, 0 GE, 0 range).
NOTE: The OPTMILP presolver value AUTOMATIC is applied.
NOTE: The OPTMILP presolver removed 0 variables and 0 constraints.
NOTE: The OPTMILP presolver removed 0 constraint coefficients.
NOTE: The OPTMILP presolver modified 0 constraint coefficients.
NOTE: The presolved problem has 15 variables, 8 constraints, and 30 constraint coefficients.
NOTE: The MIXED INTEGER LINEAR solver is called.
          Node  Active    Sols    BestInteger      BestBound        Gap     Time
             0       1       1         165000         165000      0.00%        0
             0       0       1         165000              .      0.00%        0
NOTE: OPTMILP added 0 cuts with 0 cut coefficients at the root.
NOTE: Optimal.
NOTE: Objective = 165000.
STATUS=OK SOLUTION_STATUS=OPTIMAL OBJECTIVE=165000 RELATIVE_GAP=0 ABSOLUTE_GAP=0
PRIMAL_INFEASIBILITY=0 BOUND_INFEASIBILITY=0 INTEGER_INFEASIBILITY=0 NODES=1 ITERATIONS=2
PRESOLVE_TIME=0 SOLUTION_TIME=0
```

7.2 PCB Component Allocation Problem

7.2.1 Concept of PCB Component Allocation Problem

After board types have been assigned to assembly lines or the line assignment problem (see Section 7.1) has been solved, components on the board should be grouped and allocated to appropriate SMT placement machines to achieve better line performance in terms of cycle time. Due to the various configurations of machines, different machines have different unit assembly times for the same type of surface mount components. Occasionally a machine cannot handle a particular type of component. In that case, the unit assembly time should be considered infinite (∞). Figure 7.8 shows n component types to be allocated to four placement machines in a line (Ho and Ji 2006b).

Suppose that m nonidentical placement machines are in an SMT assembly line and a board with n types of surface mount components is going to be assembled on that line. It requires t_{ij} time per unit to place if component type j is assigned to machine i. By introducing s_i to denote the setup time for machine i and c_j for the quantity of component type j, this component allocation problem with the objective of minimizing the cycle time can be formulated as shown in Model 7.2.1 (Ho and Ji 2006b).

Model 7.2.1   Standard PCB component allocation model

$$\text{Minimize } z = \max\left(s_i + \sum_{j=1}^{n} t_{ij}x_{ij} \right) \quad \text{for } i = 1, 2, \ldots, m \qquad (7.2.1)$$

subject to

$$\sum_{i=1}^{m} x_{ij} = c_j \quad \text{for } j = 1, 2, \ldots, n \qquad (7.2.2)$$

$x_{ij} \geq 0$ and is a set of integers

Decision variables x_{ij} are introduced to indicate the number of component type j to be assembled on machine i. Objective function 7.2.1 minimizes the assembly time for the machine with the largest assembly time, including the *machine setup time*, or the *cycle time*, which is defined as the maximum

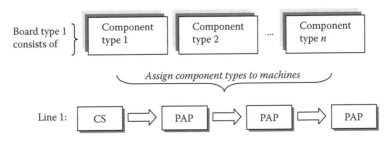

FIGURE 7.8
An example of PCB component allocation problem. CS, chip shooter; PAP, pick and place.

assembly time among all the placement machines in a line. Constraint set 7.2.2 ensures that all the components will be assembled.

Model 7.2.1 is a minimax-type integer linear programming model. To convert it into a conventional minimization type formulation, variable T is introduced.

Model 7.2.2 Transformed PCB component allocation model

$$\text{Minimize } z = T \tag{7.2.3}$$

subject to

$$T - \left(s_i + \sum_{j=1}^{n} t_{ij} x_{ij} \right) \geq 0 \quad \text{for } i = 1, 2, \ldots, m \tag{7.2.4}$$

$$\sum_{i=1}^{m} x_{ij} = c_j \quad \text{for } j = 1, 2, \ldots, n \tag{7.2.5}$$

$$x_{ij} \geq 0 \text{ and is a set of integers}$$

Constraint set 7.2.4 ensures that T will be greater than or equal to the total setup and assembly times of each placement machine. Therefore, minimization of T in objective function 7.2.3 is equivalent to the minimization of the cycle time.

7.2.2 Example of PCB Component Allocation Problem

Table 7.3 shows a component allocation problem tableau, which has three SMT placement machines and seven types of components (Ho and Ji 2006b).

TABLE 7.3

A PCB Component Allocation Problem Tableau

Machine i	Component type j							s_i
	1	2	3	4	5	6	7	
1	3 x_{11}	7 x_{12}	7 x_{13}	5 x_{14}	∞ x_{15}	∞ x_{16}	∞ x_{17}	110
2	7 x_{21}	12 x_{22}	15 x_{23}	16 x_{24}	15 x_{25}	15 x_{26}	21 x_{27}	147
3	23 x_{31}	38 x_{32}	35 x_{33}	35 x_{34}	27 x_{35}	33 x_{36}	43 x_{37}	147
c_j	324	37	12	5	7	5	4	

The upper-right corner in cell (i, j) indicates the unit assembly time t_{ij}. (The time unit is 0.1 second in the table.) For example, in cell $(1, 1)$, t_{11} is 0.3 second. If a machine cannot handle a particular type of component, the unit assembly time considered ∞.

By introducing decision variables x_{ij} to represent the amount of component type j to be allocated to machine i, this component allocation problem can be formulated as shown in Model 7.2.3.

Model 7.2.3 Example of formulation of PCB component allocation problem

$$\text{Minimize } z = T \qquad (7.2.6)$$

subject to

$$T - 3\,x_{11} - 7\,x_{12} - 7\,x_{13} - 5\,x_{14} - 9999\,x_{15} - 9999\,x_{16} - 9999\,x_{17} \geq 110 \qquad (7.2.7)$$

$$T - 7\,x_{21} - 12\,x_{22} - 15\,x_{23} - 16\,x_{24} - 15\,x_{25} - 15\,x_{26} - 21\,x_{27} \geq 147 \qquad (7.2.8)$$

$$T - 23\,x_{31} - 38\,x_{32} - 35\,x_{33} - 35\,x_{34} - 27\,x_{35} - 33\,x_{36} - 43\,x_{37} \geq 147 \qquad (7.2.9)$$

$$x_{11} + x_{21} + x_{31} = 324 \qquad (7.2.10)$$

$$x_{12} + x_{22} + x_{32} = 37 \qquad (7.2.11)$$

$$x_{13} + x_{23} + x_{33} = 12 \qquad (7.2.12)$$

$$x_{14} + x_{24} + x_{34} = 5 \qquad (7.2.13)$$

$$x_{15} + x_{25} + x_{35} = 7 \qquad (7.2.14)$$

$$x_{16} + x_{26} + x_{36} = 5 \qquad (7.2.15)$$

$$x_{17} + x_{27} + x_{37} = 4 \qquad (7.2.16)$$

$$x_{ij} \geq 0 \text{ and is a set of integers}$$

Constraint sets 7.2.7 to 7.2.9 calculate the total setup and assembly times of each placement machine. For example, constraint set 7.2.7 calculates the times for machine 1. Constraint sets 7.2.10 to 7.2.16 are the requirement constraints. There is one such constraint for each component type. For example, constraint set 7.2.10 ensures that the total amount of component type 1 must be allocated to the placement machine or machines.

7.2.3 ORCAP: SAS Code for PCB Component Allocation Problem

ORCAP is a macro that solves component allocation problems, the objective of which is to minimize the total cycle time by allocating the best combination of components to each machine (see program "sasor_7_2.sas"). The primary procedure used for the component allocation problem is PROC OPTMODEL.

Figure 7.9 illustrates the data flow in the ORCAP. It shows:

- The time matrix that is required for ORCAP, in which the unit time of any machine *i* and component *j* are specified
- The macros (%data, %model, and %report)
- The results datasets that are available for print or can be used for further analysis

In the rest of this section, the procedure of solving the PCB component allocation problem (ORCAP) in SAS, together with an example, is explained. ORCAP runs three macros: data-handling (%data), model-building (%model), and report-writing (%report).

7.2.4 ORCAP: Data-Handling Macro (%data)

This part of ORCAP processes the data into a format that is suitable for PROC OPTMODEL. The ORCAP requires only one data set containing names of machines and components and the time matrix and demand. The data set should be a .txt file, which is saved as "text tab delimited." The machines' and components' names must start with a letter and may contain up to 50 characters. The components' names must be listed in the first row of the data file. The machines' names should be listed in the first column. The number of machines and the number of components should be given to the macro

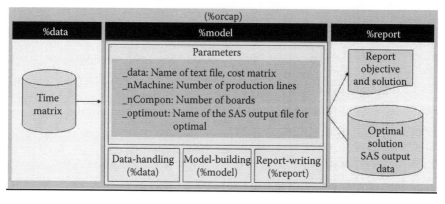

FIGURE 7.9
Data flow in ORCAP.

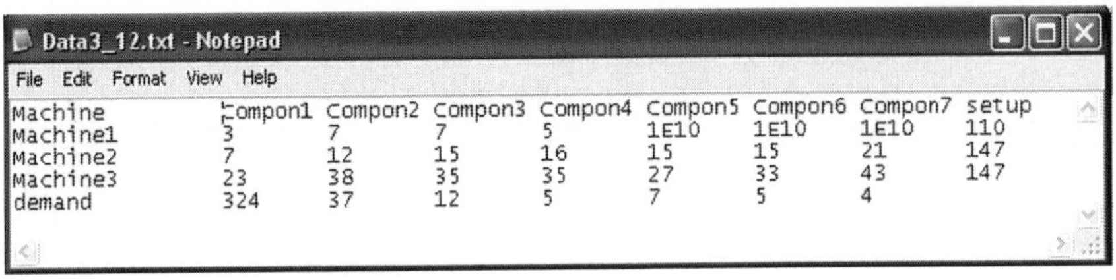

Data3_12.txt - Notepad

Machine	Compon1	Compon2	Compon3	Compon4	Compon5	Compon6	Compon7	setup
Machine1	3	7	7	5	1E10	1E10	1E10	110
Machine2	7	12	15	16	15	15	21	147
Machine3	23	38	35	35	27	33	43	147
demand	324	37	12	5	7	5	4	

FIGURE 7.10
An example of a dataset, components, machines, demand, and time matrix.

before calling it. The demand for each component should be listed in the last row of the dataset. An example data file is shown in Figure 7.10.

One parameter needs to be set before calling the data macro:

_data: Indicates the name and location of the data file (a text tab delimited file) and contains the time matrix

```
* The data-handling macro;
%macro data;
* Import text tab delimited data file to SAS data file;
proc import
      datafile = &_data
      out = data12
      dbms = tab
      replace;
      getnames = yes;
 run;
%mend data;
```

7.2.5 ORCAP: Model-Building Macro (%model)

This part of ORCAP calls PROC OPTMODEL to solve the model. Two parameters need to be set before calling the model macro:

_nMachine: Identifies the number of machines

_nCompon: Identifies the number of components

The SAS macro for model-building is as follows:

```
* The model-building macro;
%macro model;

* Starting OPTMODEL Procedure;
proc optmodel;
```

```
* Define sets;
set MACHINES = 1..&_nMachine;
set COMPONENTS = 1..&_nCompon;

* Define variables;
var X{MACHINES, COMPONENTS} integer > = 0;
var T;

* Define parameters;
number time{MACHINES, COMPONENTS};
number s{MACHINES};
number c{COMPONENTS};

* Load the time matrix;
read data data12 (where = (machine ne "demand"))
into [_N_]
{j in COMPONENTS} < time[_N_,j] = col('compon'||j) > ;

* Load the component array;
read data data12 (where = (machine eq "demand"))
into
{j in COMPONENTS} < c[j] = col('compon'||j) > ;

print c;

* Load the setup-time array;
read data data12 (where = (machine ne "demand"))
into [_N_] s[_N_] = col("setup");

* Define objective function;
min obj = T;

* Define constrains;
con Total{i in MACHINES}:
T- s[i] - sum{j in COMPONENTS: time[i,j] ne 1E10}
time[i,j]*X[i,j] > = 0;
con compon{j in COMPONENTS: j ne 0}: sum{i in MACHINES}
X[i,j] = c[j];

con zero{i in MACHINES, j in COMPONENTS :
time[i,j] = 1E10}: x[i,j] = 0;

* Solve the model;
solve with milp;
%put &_OROPTMODEL_;

expand;

* Create optimum values in a SAS dataset 'optimout';
create data &_optimout
from [MACHINES COMPONENTS]
 = {i in MACHINES, j in COMPONENTS: x[i,j]^ = 0}
```

```
amount = x ;
* End of OPTMODEL Procedure;
quit;
%mend model;
```

7.2.6 ORCAP: Report-Writing Macro (%report)

The outputs from ORCAP include one report in the form of a table of machines and components. This information is also saved in the "optimout" dataset. The user can define appropriate names for the datasets before calling the %orcap macro:

_optimout: Identifies the name of the SAS output file for "optimal solution"

Another parameter needs to be set before calling this macro:

_title: Gives a title in the output of the SAS

```
* The report-writing macro;
%macro report;
* Report the results in a tabulated form;
proc tabulate data = &_optimout;
title &_title;
class MACHINES COMPONENTS ;
var amount;
table MACHINES = " Machine",
      COMPONENTS*amount*sum
      / BOX = 'Machines or the Component Allocation
Problem';
run;
%mend report;
```

7.2.7 ORCAP: Macro (%orcap)

To make the system as user friendly as possible, the %orcap macro combines the data-handling, model-building, and report-writing codes.

```
* A SAS macro for SMT placement machines or the component
allocation problem;
%macro orcap;
 %data;
 %model;
 %report;
%mend orcap;
```

In this code, the %orcap macro is used to manage all the codes explained earlier, including data-handling, model-building, and report-writing. To get the results, the user needs to set up the parameters and run only one statement:

```
%orcap;
```

7.2.8 Instruction for Using ORCAP Macro

This section presents SAS code for the earlier example of the PCB component allocation problem with three machines and seven component types as shown in Table 7.3. The data are saved in files "data7_2.txt".

Data7_2.txt - Notepad									
File Edit Format View Help									
Machine	Compon1	Compon2	Compon3	Compon4	Compon5	Compon6	Compon7	setup	
Machine1	3	7	7	5	1E10	1E10	1E10	110	
Machine2	7	12	15	16	15	15	21	147	
Machine3	23	38	35	35	27	33	43	147	
demand	β24	37	12	5	7	5	4		

The user needs to set the parameters as required and run the following code:

```
option nodate ;
option nonumber ;
%let _title = 'Example 7.2: SMT placement machines or the
component allocation problem using PROC OPTMODEL';
%let _data = 'c:/sasor/Data7_2.txt';
%let _nMachine = 3;
%let _nCompon = 7;
%let _optimout = mysolution;
%orcap;
```

This code determines the results based on the specified parameters and the time matrix saved in the text files; it also produces a macro variable (_OROPTMODEL_) at termination. Users can examine the results of this macro variable, examine whether PROC OPTMODEL ran correctly, and examine what error or difficulty it encountered. A summary of information, including the objective value at optimum level and the status of _OROPTMODEL_, can be seen in the log file as shown in Figure 7.11.

7.2.9 Sample Results from ORCAP Macro: Output from SAS

The results of running this code are presented in Figures 7.11 and 7.12, which show the results of the primal model and the dual model,

```
Log - (Untitled)                                                                    _□×
NOTE: The MIXED INTEGER LINEAR solver is called.
      Node  Active    Sols    BestInteger        BestBound        Gap      Time
        0      1        0                        968.2405063        .         0
        0      1        1      973.0000000       969.4166666      0.37%       0
        0      1        1      973.0000000       969.9772404      0.31%       0
        0      1        1      973.0000000       970.0029009      0.31%       0
        0      1        2      971.0000000       970.0069444      0.10%       0
        0      1        2      971.0000000       970.0085580      0.10%       0
NOTE: OPTMILP added 5 cuts with 95 cut coefficients at the root.
        7      7        3      971.0000000       970.0085580      0.10%       0
      100     32        3      971.0000000       970.2011842      0.08%       0
      162     42        5      971.0000000       970.4451534      0.06%       0
      200     39        5      971.0000000       970.5325080      0.05%       0
      300     40        5      971.0000000       970.6129743      0.04%       0
      387      1        5      971.0000000       970.9387755      0.01%       0
NOTE: Optimal within relative gap.
NOTE: Objective = 971.
STATUS=OK SOLUTION_STATUS=OPTIMAL_RGAP OBJECTIVE=971 RELATIVE_GAP=0.000063057
ABSOLUTE_GAP=0.0612245173 PRIMAL_INFEASIBILITY=1.024206E-15 BOUND_INFEASIBILITY=0
INTEGER_INFEASIBILITY=1.421085E-14 NODES=388 ITERATIONS=1123 PRESOLVE_TIME=0 SOLUTION_TIME=0.265
```

FIGURE 7.11
Results of %ORCAP, solution of dual program.

Machines or the Component Allocation Problem	COMPONENTS						
	1	2	3	4	5	6	7
	amount	amount	amount	amount	amount	amount	amount
	Sum	Sum	Sum	Sum	Sum	Sum	Sum
Machine							
1	274.00	.	2.00	5.00	.	.	.
2	50.00	37.00	.	.	1.00	1.00	.
3	.	.	10.00	.	6.00	4.00	4.00

FIGURE 7.12
Results of %ORCAP, optimal solution.

respectively. In Figure 7.12, there are 21 decision variables, starting from Compon1 to Compon21. Compon1 is equivalent to x_{11} in Model 7.2.3, whereas Compon8 is equivalent to x_{21}. The optimal solution is $x_{11} = 274$, $x_{13} = 2$, $x_{14} = 5$, $x_{21} = 50$, $x_{22} = 37$, $x_{25} = 1$, $x_{26} = 1$, $x_{33} = 10$, $x_{35} = 6$, $x_{36} = 4$, and $x_{37} = 4$. The optimal solution value is 971—that is, 97.1 seconds. Figure 7.11 also shows the output from the SAS log file. The total computational time is 0.046 seconds.

7.2.10 Exercise

Use the codes developed in this chapter and solve the component assignment problem in Table 7.4.
Solution:

- Create the data in a text file (see "data7_2_exercise.txt").
- Run the following code (see program "sasor_7_2_exercise.sas"):

TABLE 7.4

A PCB Component Allocation Exercise

Machine i	Component type j							s_i
	1	2	3	4	5	6	7	
1	5 x_{11}	8 x_{12}	10 x_{13}	12 x_{14}	∞ x_{15}	∞ x_{16}	∞ x_{17}	120
2	14 x_{21}	18 x_{22}	18 x_{23}	19 x_{24}	20 x_{25}	24 x_{26}	25 x_{27}	150
3	22 x_{31}	25 x_{32}	27 x_{33}	28 x_{34}	28 x_{35}	30 x_{36}	30 x_{37}	160
4	25 x_{41}	26 x_{42}	28 x_{43}	30 x_{44}	35 x_{45}	37 x_{46}	38 x_{47}	170
c_j	500	100	80	50	30	20	10	

```
* SAS macro for SMT placement machines or the component
allocation problem: solution to exercise 7.2.;
%let _title = 'SMT placement machines or the component
allocation problem, solution to exercise 7.2';
%let _data = 'c:/sasor/Data7_2_exercise.txt';
%let _nMachine = 4;
%let _nCompon = 7;
%let _optimout = mysolution;
%orcap;
```

The following solution is given by SAS:

Machines or the Component Allocation Problem	COMPONENTS						
	1	2	3	4	5	6	7
	amount	amount	amount	amount	amount	amount	amount
	Sum	Sum	Sum	Sum	Sum	Sum	Sum
Machine							
1	482.00	.	.	.	.	.	.
2	18.00	.	79.00	34.00	3.00	.	.
3	.	14.00	.	14.00	26.00	20.00	10.00
4	.	86.00	1.00	2.00	1.00	.	−0.00

7.3 PCB Component-Sequencing Problem for PAP Machines

7.3.1 Concept of PCB Component-Sequencing Problem

After component types have been allocated to SMT placement machines or the component allocation problem (see Section 7.2) has been solved, the sequence of component placements within a machine should be determined to achieve better machine performance in terms of placement time.

The PAP machine can achieve high accuracy and is suitable for operating with large components, such as integrated circuits. Fuji XP-241E machine belongs to the class of PAP machines. In this type of placement machine, an image camera is installed on the placement head. The head can therefore move directly from the pickup points (i.e., the stationary feeders) to the placement points (i.e., the position of components on the PCB) without a stop for part image acquisition (Ho and Ji 2004; 2005; 2006b).

The operating sequence of the PAP machine is described here. First, the placement head starts from its original location (i.e., starting point), moves to a feeder that carries components, and picks up a component from the feeder. Then it moves to the desired placement location on the stationary board and places it there. After that, the head moves either back to the previous feeder if the next component is the same type as the previous one or to another feeder if it is different from the previous one, where it picks up the next component and repeats the operating procedure. After completing all component placements on a board, the head returns to its original location and waits for the next board to be assembled, as shown in Figure 7.13 for 10 components (Ho and Ji 2006b).

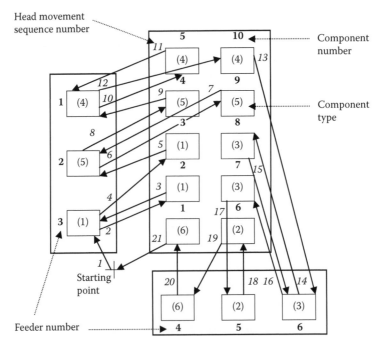

FIGURE 7.13
The assembly sequence of the placement head.

Consider a board with 10 components of 6 types that requires assembly using the PAP machine, as illustrated in Figure 7.13. The number inside the bracket represents the component type. For example, component 1 or c_1 is of type 6. Furthermore, each of the component types is assigned to a feeder. For instance, component type 4 is stored in feeder 1, or f_1. If the sequence of placements starts with component 2, then components 3, 9, 4, 5, 10, 8, 7, 6, and finally component 1, then the entire assembly sequence of the placement head will be starting point $\to f_3 \to c_2 \to f_3 \to c_3 \to f_2 \to c_9 \to f_2 \to c_4 \to f_1 \to c_5 \to f_1 \to c_{10} \to f_6 \to c_8 \to f_6 \to c_7 \to f_5 \to c_6 \to f_4 \to c_1 \to$ starting point.

Suppose that the assignment of component types to feeders (i.e., the feeder arrangement problem) is solved beforehand. Then the component-sequencing model can be formulated to find the minimal distance traveled by the placement head for assembling all components on the PCB. To achieve this goal, a decision variable is defined as

$$x_{ij} = \begin{cases} 1 & \text{if component } i \text{ is placed immediately prior to component } j \\ 0 & \text{otherwise} \end{cases}$$

Actually, the *component-sequencing problem* is somewhat similar to the traveling salesman problem (TSP), discussed in Section 6.1, except for the objective function. For the TSP, the objective is simply to minimize $\sum_{i=1}^{n} \sum_{\substack{j=1 \\ j \neq i}}^{n} c_{ij} x_{ij}$, where c_{ij} is the distance between cities i and j. For the PAP machine, the objective is not to minimize the distance between components i and j, because the placement head is unable to place the next component on the PCB immediately without picking up a component from a feeder first. Therefore, the objective for the PAP machine should be to minimize the summation of different distances, including:

- The distance between the position of component i on the PCB and feeder l (if $i = 0$, it is the distance between the starting point at the beginning and feeder l)
- The distance between feeder l and the position of the next component j
- The distance between the position of the last component i and the starting point at the end

For example, if the sequence of component placements starts with component 2 and finishes with component 1, as shown in Figure 7.13, then both decision variables x_{02} and x_{10} are equal to 1. As mentioned before, it is assumed that the feeder arrangement problem is solved beforehand. If component 2 is stored in feeder 3, then the placement head travels from the starting point to feeder 3 initially to pick up a component and then moves from feeder 3 to the position of component 2 to place the component. So the distances for assembling component 2 include the distance from the starting point to feeder 3 (i.e., $d_{il} = d_{03}$) and the distance from feeder 3 to the position of component 2 (i.e., $d_{lj} = d_{32}$). The idea of calculating the distances for assembling the remaining $(n - 1)$ components is the same. In addition, the distance for the placement head to return from the position of the last component to the starting point should be included (i.e., $d_{i0} = d_{10}$). The mathematical model for the component-sequencing problem can be formulated as shown in Model 7.3.1 (Ho and Ji 2006b).

Model 7.3.1   Standard PCB component-sequencing model

$$\text{Minimize } z = \sum_{i=0}^{n} \sum_{\substack{j=1 \\ j \neq i}}^{n} \sum_{l=1}^{\mu} \left(d_{ij} + d_{ij} \right) x_{ij} + \sum_{i=1}^{n} d_{i0} x_{i0} \qquad (7.3.1)$$

subject to

$$\sum_{i=0}^{n} x_{ij} = 1 \quad \text{for } j = 0, 1, \ldots, n; \, i \neq j \qquad (7.3.2)$$

$$\sum_{j=0}^{n} x_{ij} = 1 \quad \text{for } i = 0, 1, \ldots, n; \, i \neq j \qquad (7.3.3)$$

$$u_i - u_j + n x_{ij} \leq n - 1 \quad \text{for } i, j = 1, 2, \ldots, n; \, i \neq j \qquad (7.3.4)$$

All $x_{ij} = 0$ or 1. All $u_i \geq 0$ and is a set of integers.

In Model 7.3.1, objective function 7.3.1 minimizes the total travel distance of the placement head. If the moving speed of the placement head is incorporated, then the objective minimizes the total placement time for assembling all components on the PCB. Constraint set 7.3.2 ensures that exactly one component is placed immediately before component j. Constraint set 7.3.3 ensures that exactly one component is placed immediately after component i. Although the solution drawn satisfies both constraint sets 7.3.2 and 7.3.3, it may still be unfeasible due to the occurrence of subtours. Therefore, constraint set 7.3.4 is added to eliminate subtours. Because the starting point must be visited first, it is redundant to include $i = 0$ or $j = 0$ (or both) in constraint set 7.3.4. This is very similar to the classic TSP except that the placement head has to pick up a component from a feeder before placing the component to its position.

7.3.2 Example of PCB Component-Sequencing Problem

Consider a board with four components of four types to be assembled by the PAP machine. The types of the components and the locations of both components and feeders are shown in Table 7.5. Because the distances are symmetric (i.e., $d_{il} = d_{li}$), the distance between a feeder and a component—whether it is d_{il} or d_{ij}—can be found in Table 7.6. Also, the distances d_{i0} are summarized in Table 7.7. Note that the calculation of the distances is based on the Euclidean metric and that the coordinates of the starting point are (0, 0).

Suppose that the assignment of component types to feeders is generated randomly so that component types 1, 3, 2, and 4 are stored in feeders 1, 2, 3, and 4, respectively. Based on this initial feeder arrangement, the component sequencing problem in the form of Model 7.3.1 becomes more involved, as shown in Model 7.3.2.

TABLE 7.5

Data of Components and Feeders

Components (i)	Types	Coordinates (mm) x	y	Feeders (l)	Coordinates (mm) x	y
1	4	30	40	1	10	30
2	3	30	60	2	10	20
3	1	50	20	3	20	10
4	2	50	40	4	30	10

TABLE 7.6

Distance between a Feeder and a Component

l \ i	0	1	2	3	4
1	31.62	22.36	36.06	41.23	41.23
2	22.36	28.28	44.72	40	44.72
3	22.36	31.62	50.99	31.62	42.43
4	31.62	30	50	22.36	36.06

TABLE 7.7

Distance between the Starting Point and a Component

i	1	2	3	4
j = 0	50	67.08	53.85	64.03

Model 7.3.2 Example of formulation of PCB component sequencing problem

$$\text{Minimize } z = 61.62\, x_{01} + 67.08\, x_{02} + 72.85\, x_{03} + 64.79\, x_{04}$$

$$+ 73.00\, x_{12} + 63.59\, x_{13} + 74.05\, x_{14}$$

$$+ 80.00\, x_{21} + 77.29\, x_{23} + 93.42\, x_{24}$$

$$+ 52.36\, x_{31} + 84.72\, x_{32} + 74.05\, x_{34}$$

$$+ 66.06\, x_{41} + 89.44\, x_{42} + 82.46\, x_{43}$$

$$+ 50.00\, x_{10} + 67.08\, x_{20} + 53.85\, x_{30} + 64.03\, x_{40} \qquad (7.3.5)$$

subject to

$$x_{10} + x_{20} + x_{30} + x_{40} = 1 \qquad\qquad (7.3.6)$$

$$x_{01} + x_{21} + x_{31} + x_{41} = 1 \qquad\qquad (7.3.7)$$

$$x_{02} + x_{12} + x_{32} + x_{42} = 1 \qquad\qquad (7.3.8)$$

$$x_{03} + x_{13} + x_{23} + x_{43} = 1 \qquad\qquad (7.3.9)$$

$$x_{04} + x_{14} + x_{24} + x_{34} = 1 \qquad\qquad (7.3.10)$$

$$x_{01} + x_{02} + x_{03} + x_{04} = 1 \qquad\qquad (7.3.11)$$

$$x_{10} + x_{12} + x_{13} + x_{14} = 1 \qquad\qquad (7.3.12)$$

$$x_{20} + x_{21} + x_{23} + x_{24} = 1 \qquad\qquad (7.3.13)$$

$$x_{30} + x_{31} + x_{32} + x_{34} = 1 \qquad\qquad (7.3.14)$$

$$x_{40} + x_{41} + x_{42} + x_{43} = 1 \qquad\qquad (7.3.15)$$

$$u_1 - u_2 + 4\, x_{12} \le 3 \qquad\qquad (7.3.16)$$

$$u_1 - u_3 + 4\, x_{13} \le 3 \qquad\qquad (7.3.17)$$

$$u_1 - u_4 + 4\, x_{14} \le 3 \qquad\qquad (7.3.18)$$

$$u_2 - u_1 + 4\, x_{21} \le 3 \qquad\qquad (7.3.19)$$

$$u_2 - u_3 + 4\,x_{23} \leq 3 \tag{7.3.20}$$

$$u_2 - u_4 + 4\,x_{24} \leq 3 \tag{7.3.21}$$

$$u_3 - u_1 + 4\,x_{31} \leq 3 \tag{7.3.22}$$

$$u_3 - u_2 + 4\,x_{32} \leq 3 \tag{7.3.23}$$

$$u_3 - u_4 + 4\,x_{34} \leq 3 \tag{7.3.24}$$

$$u_4 - u_1 + 4\,x_{41} \leq 3 \tag{7.3.25}$$

$$u_4 - u_2 + 4\,x_{42} \leq 3 \tag{7.3.26}$$

$$u_4 - u_3 + 4\,x_{43} \leq 3 \tag{7.3.27}$$

All $x_{ij} = 0$ or 1. All $u_i \geq 0$ and is a set of integers.

Constraint sets 7.3.6 to 7.3.10 ensure that exactly one component is placed immediately before component j. There is one such constraint for each component. For example, constraint set 7.3.6 ensures that exactly one component is placed immediately before returning to the starting point. Constraint sets 7.3.11 to 7.3.15 make sure that exactly one component is placed immediately after component i. Similarly, there is one such constraint for each component. For example, constraint set 7.3.11 ensures that exactly one component is placed just after the starting point at the beginning. Constraint sets 7.3.16 to 7.3.27 are the subtour elimination constraints.

7.3.3 ORCSP: SAS Code for PCB Component-Sequencing Problem

ORCSP is a macro that solves component-sequencing problems, the objective of which is to minimize the total placement time (see program "sasor_7_3. sas"). The primary procedure used for the component-sequencing problem is PROC LP with integer programming.

Figure 7.14 illustrates the data flow in the ORCSP. It shows:

- The coordinates of components and the coordinates of feeders that are required for ORCSP
- The macros (%data, %model, and %report)
- The results datasets that are available for print or can be used for further analysis

In the rest of this section, the procedure for implementing the PCB component-sequencing problem (ORCSP) in SAS, together with an example, is

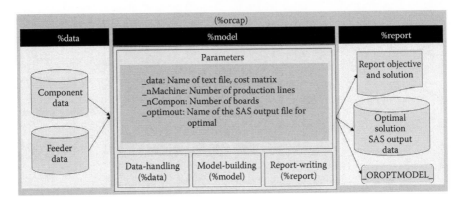

FIGURE 7.14
Data flow in ORCSP.

FIGURE 7.15
An example of a dataset, components' information.

explained. ORCSP runs three macros: data-handling (%data), model-building (%model), and report-writing (%report).

7.3.4 ORCSP: Data-Handling Macro (%data)

This part of ORCSP processes the data into a format suitable for PROC LP. The ORCSP requires two datasets containing information about components, including type of components and coordinates of components, and information about feeders, including coordinates of randomly assigned feeders. Example data files are shown in Figures 7.15 and 7.16.

Two parameters need to be set before calling the data macro:

_dataC: Indicates the name and location of the component data file (a text tab delimited file) and contains components' information

_dataF: Indicates the name and location of the feeder data file (a text tab delimited file) and contains feeders' information

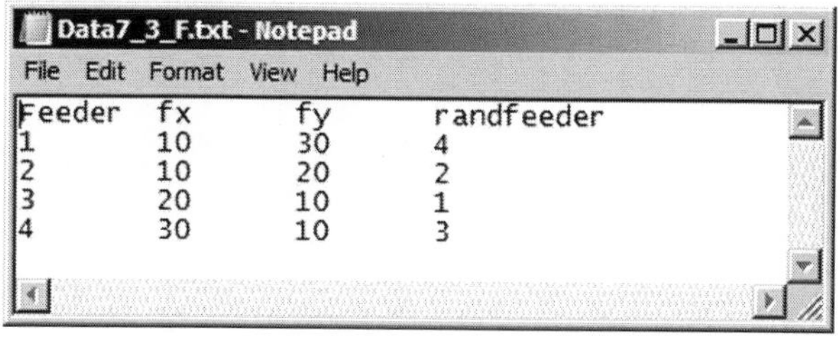

Data7_3_F.txt - Notepad

File Edit Format View Help

Feeder	fx	fy	randfeeder
1	10	30	4
2	10	20	2
3	20	10	1
4	30	10	3

FIGURE 7.16
An example of a dataset, feeders' information.

```
* The data-handling macro;
%macro data;
* Import text tab delimited data file to SAS data file;
proc import
 datafile = &_dataC
 out = dataC
 dbms = tab
 replace;
 getnames = yes;
run;
proc import
 datafile = &_dataF
 out = dataF
 dbms = tab
 replace;
 getnames = yes;
 run;
%mend data;
```

7.3.5 ORCSP: Model-Building Macro (%model)

This part of ORCSP calls PROC OPTMODEL to solve the model. Two parameters need to be set before calling the model macro:

_nFeeder: Identifies the number of feeders

_nCompon: Identifies the number of components

The SAS macro for model-building is as follows:

```
* The model-building macro;
%macro model;
* Starting OPTMODEL Procedure;
proc optmodel;
```

```
* Define sets;
set COMP = 0..&_nCompon;
set Feed = 0..&_nFeeder;

* Define variables;
var u{COMP} integer > = 0;
var x{COMP,COMP} binary > = 0 ;
var z;

* Define parameters;
number d{COMP,COMP};
number cx {COMP};
number cy {COMP};
number type {COMP};
number fx {FEED};
number fy {FEED};
number randfeeder {FEED};
number d1;
number d2;
number Distance{COMP,FEED};
number ds2{COMP,FEED};

* Load component data;
read data Datac into [_N_] type[_N_] = col('type');
read data Datac into [_N_] cx[_N_] = col('cx');
read data Datac into [_N_] cy[_N_] = col('cy');

* Load feeder data;
read data Dataf into [_N_] fx[_N_] = col('fx');
read data Dataf into [_N_] fy[_N_] = col('fy');
read data Dataf into [_N_] randfeeder[_N_] =
col('randfeeder');

cx[0] = 0; cy[0] = 0; fx[0] = 0; fy[0] = 0;

for {i in FEED, j in COMP}
do;
 if (i ne 0 & j ne 0)
      then do; d1 = (cx[i]-fx[j])**2; d2 = (cy[i]-
      fy[j])**2; end;
      else if (i = 0 & j = 0)
            then do; d1 = 0; d2 = 0; end;
            else if (i = 0)
                  then do; d1 = (cx[j]**2); d2 =
                  (cy[j]**2); end;
                  else if (j = 0)
                  then do; d1 = (fx[i]**2); d2 =
                  (fy[i]**2); end;
 Distance[i,j] = round(sqrt( d1 + d2),0.000001);
end;
```

```
print Distance;
for {i in FEED, j in COMP}
do;
     if (i = j) then ds2[i,j] = 0;
     else if i = 0 then
ds2[j,i] = Distance[randfeeder[j],0]+Distance[j,
randfeeder[j]];
     else if j = 0 then ds2[j,i] = Distance[j,i];
     else ds2[j,i] = Distance[j,randfeeder[j]] +
     Distance[i,randfeeder[j]];
end;

* Define objective function;
min obj = z;

* Define constrains;
con Objective: z=sum{i in FEED, j in COMP}
ds2[j,i]*x[i,j];

con component_immediately_before1 {i in FEED}: sum{j in
COMP: i ne j} x[j,i] = 1;

con component_immediately_before2 {j in COMP}: sum{i in
FEED: i ne j} x[j,i] = 1;
con subtour_elimination {i in FEED, j in COMP : i ne 0 &
j ne 0 & i ne j}: u[i]-u[j] + &_nCompon * x[i,j] < =
&_nCompon-1;
solve with MILP;

expand;

%put &_OROPTMODEL_;

* Create optimum values of x in a SAS dataset
'optimout1';
create data optimout1
from [COMP FEED]
 = {i in COMP, j in FEED: x[i,j] ne 0}
amount = x ;

* Create optimum values of u in a SAS dataset
'optimout2';
create data optimout2
from [COMP]
= {i in COMP: u[i]}
amount = u ;

*End of PROC OPTMODE;
quit;
%mend model;
```

7.3.6 ORCSP: Report-Writing Macro (%report)

The outputs from ORCSP include two reports. Report 1 contains all the information and the solution of the primal model that is saved in the "outprimal" dataset, and report 2 contains all the information and the solution of the dual model that is saved in the "outdual" dataset. Another dataset ("outtable"), which is created at termination, contains all information from the model in a tabulated format. The users can define appropriate names for each of these datasets before calling %orcsp macro:

_outprimal: Identifies the name of the SAS output file for "primal" information

Another parameter needs to be set before calling this macro:

_title: Gives a title in the output of the SAS

```
* The report-writing macro;
%macro report;
title &_title;

proc tabulate data = optimout1;
title &_title;
class COMP FEED ;
var amount;
table COMP = "COMP",
FEED*amount*sum
/ BOX = 'Component sequencing';
run;

proc tabulate data = optimout2;
title &_title;
class COMP ;
var amount;
table COMP = "COMP",
amount*sum
/ BOX = 'Component sequencing';
run;
%mend report;
```

7.3.7 ORCSP: Macro (%orcsp)

To make the system as user friendly as possible, the %orcsp macro combines the data-handling, model-building, and report-writing codes.

```
* A SAS macro for SMT placement machines or the component
sequencing problem;
%macro orcsp;
```

```
%data;
 %model;
 %report;
%mend orcsp;
```

In this code, the %orcsp macro is used to manage all the codes explained earlier, including data-handling, model-building, and report-writing. To get the results, user needs to set up the parameters and run only one statement: %orcsp;

7.3.8 Instructions for Using ORCSP Macro

This section presents the SAS code for the earlier example of the PCB component-sequencing problem with four components and four feeders as shown in Table 7.1. The data are saved in files "data7_3_C.txt" and "data7_3_F.txt."

The user needs to set the parameters as required and run the following code:

```
* A SAS procedure for SMT placement machines or the
component sequencing problem;
%let _title = 'Example 7.3: SMT placement machines or the
component sequencing problem';
%let _dataC = 'c:/sasor/Data7_3_C.txt';
%let _dataF = 'c:/sasor/Data7_3_F.txt';
%let _nFeeder = 4;
%let _nCompon = 4;
option nodate;
%orcsp;
```

This code determines the results based on the specified parameters and the components and feeders saved in the text files; it also produces a macro variable (_OROPTMODEL) at termination. The user can examine the result of this macro variable, examine whether PROC OPTMODEL ran correctly, and examine what error or difficulty it encountered.

7.3.9 Sample Results from ORCSP Macro: Output from SAS

The results of running this code are presented in Figure 7.17. Because it is assumed that component types 1, 3, 2, and 4 are stored in feeders 1, 2, 3, and 4, respectively, the entire assembly sequence of the placement head will be starting point $\rightarrow f_3 \rightarrow c_4 \rightarrow f_2 \rightarrow c_2 \rightarrow f_1 \rightarrow c_3 \rightarrow f_4 \rightarrow c_1 \rightarrow$ starting point. The optimal solution value is 333.88 mm. Figure 7.18 shows the output from the SAS log file.

	FEED				
	0	1	2	3	4
Component	amount	amount	amount	amount	amount
sequencing	Sum	Sum	Sum	Sum	Sum
COMP					
0	.	.	−0.00	.	1.00
1	1.00	.	0.00	.	.
2	.	.	.	1.00	.
3	.	1.00	.	.	−0.00
4	0.00	−0.00	1.00	.	.

FIGURE 7.17
Results of %ORCSP, sample output from SAS, solution of OPTMODEL program.

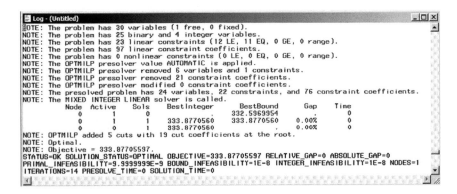

FIGURE 7.18
Log for %ORCSP.

7.3.10 Exercise

Use the codes developed in this chapter and solve the PCB component sequence problem in Tables 7.8 to 7.10. Note that the calculation of the distances is based on the Euclidean metric, and that the coordinates of the starting point are (0, 0). It is assumed that the assignment of component types to feeders is generated randomly and that component types 1, 3, 5, 2, 4, and 6 are stored in feeders 1, 2, 3, 4, 5, and 6 respectively.
Solution:

- Create the data in a text file (see "data7_3_C_exercise.txt" and "data7_3_F_exercise.txt").
- Run the following code (see program "sasor_7_3_exercise.sas"):

TABLE 7.8

Data of Components and Feeders

Components (i)	Types	Coordinates (mm) x	y	Feeders (l)	Coordinates (mm) x	y
1	4	30	40	1	10	30
2	3	30	60	2	10	20
3	5	30	80	3	10	10
4	1	50	20	4	20	10
5	2	50	40	5	30	10
6	6	50	60	6	40	10

TABLE 7.9

Distance between a Feeder and a Component

l \ i	0	1	2	3	4	5	6
1	31.62	22.36	36.06	53.85	41.23	41.23	50
2	22.36	28.28	44.72	63.25	40	44.72	56.57
3	14.14	36.06	53.85	72.80	41.23	50	64.03
4	22.36	31.62	50.99	70.71	31.62	42.43	58.31
5	31.62	30	50	70	22.36	36.06	53.85
6	41.23	31.62	50.99	70.71	14.14	31.62	50.99

TABLE 7.10

Distance between the Starting Point and a Component

i	1	2	3	4	5	6
j = 0	50	67.08	85.44	53.85	64.03	78.10

Component sequencing	FEED						
	0	1	2	3	4	5	6
	amount	amount	amount	amount	amount	amount	amount
	Sum	Sum	Sum	Sum	Sum	Sum	Sum
COMP							
0	.	.	.	.	.	1.00	.
1	1.00	.	.	.	.	.	.
2	.	.	.	1.00	.	.	.
3	.	1.00	.	.	.	.	.
4	.	.	.	.	.	.	1.00
5	.	.	.	.	1.00	.	.
6	.	.	1.00	.	.	.	.

FIGURE 7.19

Results of %ORCSP, sample output from SAS, solution of OPTMODEL program.

```
Log - (Untitled)                                                                    _ □ ×
NOTE: The problem has 56 variables (1 free, 0 fixed).
NOTE: The problem has 49 binary and 6 integer variables.
NOTE: The problem has 45 linear constraints (30 LE, 15 EQ, 0 GE, 0 range).
NOTE: The problem has 217 linear constraint coefficients.
NOTE: The problem has 0 nonlinear constraints (0 LE, 0 EQ, 0 GE, 0 range).
NOTE: The OPTMILP presolver value AUTOMATIC is applied.
NOTE: The OPTMILP presolver removed 8 variables and 1 constraints.
NOTE: The OPTMILP presolver removed 43 constraint coefficients.
NOTE: The OPTMILP presolver modified 0 constraint coefficients.
NOTE: The presolved problem has 48 variables, 44 constraints, and 174 constraint coefficients.
NOTE: The MIXED INTEGER LINEAR solver is called.
          Node  Active   Sols   BestInteger      BestBound       Gap    Time
             0       1      1    531.1696150    531.1696150     0.00%      0
             0       0      1    531.1696150               .    0.00%      0
NOTE: OPTMILP added 0 cuts with 0 cut coefficients at the root.
NOTE: Optimal.
NOTE: Objective = 531.169615.
STATUS=OK SOLUTION_STATUS=OPTIMAL OBJECTIVE=531.169615 RELATIVE_GAP=0 ABSOLUTE_GAP=0
PRIMAL_INFEASIBILITY=0 BOUND_INFEASIBILITY=0 INTEGER_INFEASIBILITY=0 NODES=1 ITERATIONS=14
PRESOLVE_TIME=0 SOLUTION_TIME=0
```

FIGURE 7.20
Log for %ORCSP.

```
* A SAS procedure for SMT placement machines or the
component sequencing problem;
%let _title = 'Example 7.3: SMT placement machines or the
component sequencing problem';
%let _dataC = 'c:/sasor/Data7_3_C_exercise.txt';
%let _dataF = 'c:/sasor/Data7_3_F_exercise.txt';
%let _nFeeder = 6;
%let _nCompon = 6;
option nodate;
%orcsp;
```

The solutions shown in Figures 7.19 and 7.20 are given by SAS.

8

Multiple-Criteria Decision Making

In this chapter, we examine multiple-criteria logistics distribution with analytic hierarchy process and demonstrate how SAS/OR® can be applied to solve the problems to optimality.

8.1 Multiple-Criteria Logistics Distribution Problem: Phase 1

8.1.1 Concept of Multiple-Criteria Logistics Distribution Problem

The *logistics distribution problem* allocates a number of points of consumption to a number of points of supply, including suppliers, manufacturers, warehouses, distribution centers, and customers. The connection of these logistics stakeholders by means of transportation facilities is regarded as the logistics distribution network. Logistics distribution network design is one of the major decision problems arising in contemporary supply chain management. There are two primary inadequacies in the traditional approaches to the problem. First, the focus is confined to the points of supply. The objective is either to minimize the total logistics cost or the total delivery time. However, consideration of the customers is neglected. Second, only quantifiable data are considered in the optimization techniques. Some qualitative factors that are mainly customer oriented are not taken into account.

To overcome these drawbacks, this section and the next section apply an integrated multiple-criteria decision-making approach that combines the *analytic hierarchy process (AHP)* and *goal-programming (GP) model* to design an optimal logistics distribution network. The integrated approach considers both quantitative and qualitative factors and also aims at maximizing the benefits to the supplier and the customer.

In the first phase of the integrated approach for the distribution network design (described in this section), the AHP is used to determine the relative importance weightings of alternative warehouses with respect to various evaluating criteria that are primarily qualitative. In the second phase (see Section 8.2), a GP model is formulated to select an optimal set of warehouses while considering the AHP priorities of warehouses and the quantitative-based limitations of resources.

8.1.2 Concept of AHP

The AHP, developed by Saaty (1980), consists of three primary operations: hierarchy construction, priority analysis, and consistency verification (Ho et al. 2006; Ho 2007). First, those using the AHP need to break down a complex multiple-criteria decision-making problems into its component parts, all possible attribute of which are arranged into multiple hierarchical levels. For example, the overall goal, criteria, and alternatives to each criterion are in the first, second, and third levels, respectively.

Second, the AHP users have to compare each cluster in the same level in a pairwise fashion based on their own experience and knowledge. For instance, every pair of criteria in the second level are compared with respect to the achieving goal while every pair of alternatives in the third level are compared with respect to their corresponding criteria. Then a judgment is made about which is more important and by how much. A nine-point scale, shown in Table 8.1, can be used to determine the priorities.

Because the comparisons are carried out through personal or subjective judgments, some degree of inconsistency may occur. To guarantee that the judgments are consistent, the final operation, called *consistency verification*, is incorporated to measure the degree of consistency among the pairwise comparisons.

The AHP is described in the following steps:

Step 1—AHP pairwise comparison: Construct a pairwise comparison matrix:

$$A = \begin{bmatrix} a_{11} & a_{12} & \cdots & a_{1n} \\ a_{21} & a_{22} & \cdots & a_{2n} \\ \vdots & \vdots & \ddots & \vdots \\ a_{n1} & a_{n2} & \cdots & a_{nn} \end{bmatrix}$$

TABLE 8.1

AHP Pairwise Comparison Scale

Intensity	Importance	Explanation
1	Equal	Two activities contribute equally to the object
3	Moderate	Slightly favors one over another
5	Strong	Strongly favors one over another
7	Very strong	Dominance of the demonstrated in practice
9	Extreme	Evidence favoring one over another of highest possible order of affirmation
2, 4, 6, 8	Intermediate	When compromise is needed
Reciprocals of the above numbers		For inverse comparison

where n denotes the number of elements (criteria or alternatives) and a_{ij} refers to the comparison of element i with element j with respect to each goal or criterion.

Step 2—AHP synthesization: Divide each entry (a_{ij}) in each column of matrix A by its column total. The matrix now becomes a normalized pairwise comparison matrix, A'.

Step 3: Compute the priorities of the elements—that is, the average of the entries in each row of matrix A'.

Step 4—AHP consistency verification: Multiply each entry in column i of matrix A by the priority of element i. Then divide the summation of values in row i by the priority of element i.

Step 5: Compute the averages of values to yield the maximum eigenvalue of matrix A, λ_{max}.

Step 6: Compute the consistency index:

$$CI = \frac{\lambda_{max} - n}{n-1}$$

Step 7: Compute the consistency ratio (CR):

$$CR = \frac{CI}{RI(n)}$$

where $RI(n)$ is a random index the value of which is dependent on the value of n, as shown in Table 8.2. If CR is greater than 0.10, the AHP users should review and revise the pairwise comparisons.

8.1.3 Example of AHP

In this section, the AHP is used to evaluate the performance of four warehouses. There are five evaluating criteria: (1) total lead time, (2) reliability of order fulfillment, (3) quality, (4) flexibility of capacity, and (5) value-added services. *Total lead time* comprises the time required to handle inventory in the warehouses, load and store inventory in the warehouses, and deliver products from the warehouses to the customers. Reliability of order fulfillment consists of delivering the correct type and quantity of product and providing on-time delivery. Quality involves the commitment of the deliverer to providing high-quality products in a condition that is acceptable to the customer. *Flexibility of capacity* refers to the ability of warehouses to respond

TABLE 8.2

List of Random Index Values

n	2	3	4	5	6	7	8	9
$RI(n)$	0	0.58	0.90	1.12	1.24	1.32	1.41	1.45

to fluctuations in the volume of customer orders. *Value-added services* refer to any activities that enhance customer satisfaction (e.g., track-and-trace, 24-hour customer hotline) and the responsiveness of the warehouses to special requests by the customer (e.g., secure packaging, urgent delivery).

The warehouse evaluation problem is shown in Figure 8.1. After constructing the hierarchy, two criteria are compared at a time with respect to achieving the goal. Once the pairwise comparisons have been made for the five criteria (Figure 8.2), each alternative warehouse is compared against each other alternative with respect to the corresponding criterion at a time (Figures 8.3 to 8.7). After completion of all pairwise comparisons, SAS is used to synthesize the relative priority of each criterion, the relative priority of each alternative, and the consistency ratios.

8.1.4 ORAHP: SAS Code for AHP

ORAHP is a macro that implements the AHP to solve the warehouse evaluation problem, the objective of which is to evaluate the performance of a set of alternative warehouses.

Figure 8.2 illustrates the data flow in ORAHP. It shows:

- The criteria and alternatives' matrices that are required for ORAHP, in which the alternatives express their preference for each criterion
- The macros (%data, %model, %consist, %priority, %report1, and %report2)
- The results datasets that are available for print or can be used for further analysis.

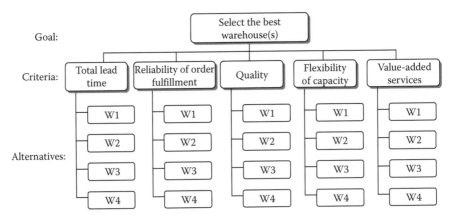

FIGURE 8.1
The hierarchy of warehouse prioritization. W1, warehouse 1; W2, warehouse 2; W3, warehouse 3; W4, warehouse 4.

$$A_1 = \begin{bmatrix} 1 & 1 & 2 & 3 & 4 \\ 1 & 1 & 2 & 3 & 4 \\ \dfrac{1}{2} & \dfrac{1}{2} & 1 & 3 & 4 \\ \dfrac{1}{3} & \dfrac{1}{3} & \dfrac{1}{3} & 1 & 1 \\ \dfrac{1}{4} & \dfrac{1}{4} & \dfrac{1}{4} & 1 & 1 \end{bmatrix}$$

FIGURE 8.2
Pairwise comparison matrix for criteria with respect to goal.

$$A_4 = \begin{bmatrix} 1 & 3 & 2 & 3 \\ \dfrac{1}{3} & 1 & \dfrac{1}{2} & 1 \\ \dfrac{1}{2} & 2 & 1 & 2 \\ \dfrac{1}{3} & 1 & \dfrac{1}{2} & 1 \end{bmatrix}$$

FIGURE 8.5
Pairwise comparison matrix for warehouses with respect to quality.

$$A_2 = \begin{bmatrix} 1 & 3 & 3 & 5 \\ \dfrac{1}{3} & 1 & 1 & 4 \\ \dfrac{1}{3} & 1 & 1 & 3 \\ \dfrac{1}{5} & \dfrac{1}{4} & \dfrac{1}{3} & 1 \end{bmatrix}$$

FIGURE 8.3
Pairwise comparison matrix for warehouses with respect to total lead time.

$$A_5 = \begin{bmatrix} 1 & \dfrac{1}{3} & \dfrac{1}{2} & \dfrac{1}{5} \\ 3 & 1 & 3 & \dfrac{1}{2} \\ 2 & \dfrac{1}{3} & 1 & \dfrac{1}{3} \\ 5 & 2 & 3 & 1 \end{bmatrix}$$

FIGURE 8.6
Pairwise comparison matrix for warehouses with respect to flexibility of capacity.

$$A_3 = \begin{bmatrix} 1 & 3 & 2 & 4 \\ \dfrac{1}{3} & 1 & \dfrac{1}{2} & 2 \\ \dfrac{1}{2} & 2 & 1 & 3 \\ \dfrac{1}{4} & \dfrac{1}{2} & \dfrac{1}{3} & 1 \end{bmatrix}$$

FIGURE 8.4
Pairwise comparison matrix for warehouses with respect to reliability of order fulfillment.

$$A_6 = \begin{bmatrix} 1 & 4 & 2 & 3 \\ \dfrac{1}{4} & 1 & \dfrac{1}{3} & 1 \\ \dfrac{1}{2} & 3 & 1 & 2 \\ \dfrac{1}{3} & 1 & \dfrac{1}{2} & 1 \end{bmatrix}$$

FIGURE 8.7
Pairwise comparison matrix for warehouses with respect to value-added services.

In the rest of this section, the procedure for implementing the AHP (ORAHP) in SAS, together with an example, is explained. ORAHP runs seven macros: data-handling (%data), model-building (%consist, %priority, and %model), and report-writing (%report1 and %report2).

8.1.5 ORAHP: Data-Handling Macro (%data)

This part of ORAHP processes the data into a format suitable for implementating AHP (see program "sasor_8_1.sas"). ORAHP requires two datasets.

The first data set contains the names of the criteria and a pairwise comparison matrix of criteria. The second dataset contains the names of the criteria in the first column and the name of the alternatives in the second column, followed by pairwise comparison matrix of the alternatives and criteria. The data sets should be .txt files, which are saved as "text tab delimited." The names of the criteria and alternatives must start with a letter and may contain up to 50 characters. Example data files are shown in Figures 8.9 and 8.10.

Two parameters need to be set before calling the data macro:

_dataC: Indicates the name and location of the data file (a text tab delimited file) and contains the criteria matrix

_dataA: Indicates the name and location of the data file (a text tab delimited file) and contains the criteria and alternatives matrix

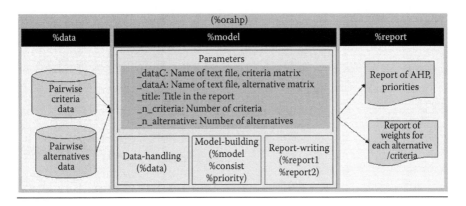

FIGURE 8.8
Data flow in ORAHP.

```
Data8_1_C - Notepad                              _ □ ×
File  Edit  Format  View  Help
Lead           1        1        2        3        4
Reliability    1        1        2        3        4
Quality                 1        3        4
Flexibility                               1        1
ValueAdded                                1        1
```

FIGURE 8.9
An example of a dataset, pairwise comparison matrix of criteria.

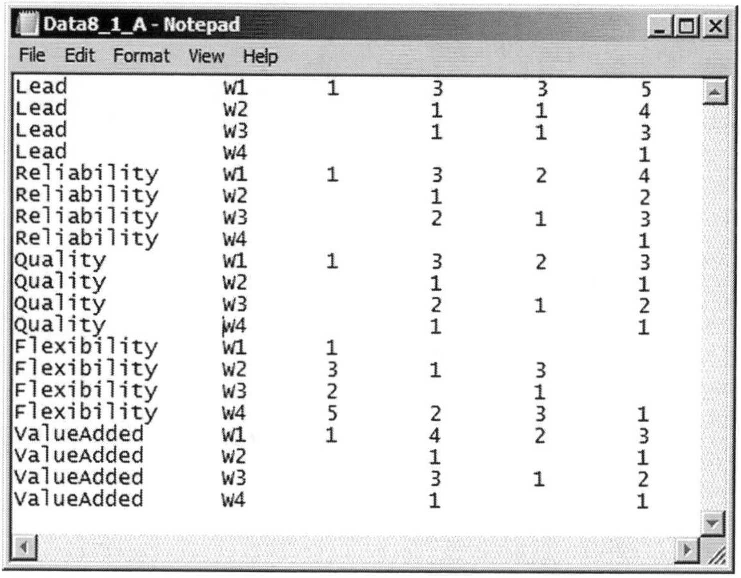

```
Data8_1_A - Notepad                              _ □ ×
File  Edit  Format  View  Help
Lead          w1     1      3      3      5
Lead          w2            1      1      4
Lead          w3            1      1      3
Lead          w4                          1
Reliability   w1     1      3      2      4
Reliability   w2            1             2
Reliability   w3            2      1      3
Reliability   w4                          1
Quality       w1     1      3      2      3
Quality       w2            1             1
Quality       w3            2      1      2
Quality       w4            1             1
Flexibility   w1     1
Flexibility   w2     3      1      3
Flexibility   w3     2             1
Flexibility   w4     5      2      3      1
ValueAdded    w1     1      4      2      3
ValueAdded    w2            1             1
ValueAdded    w3            3      1      2
ValueAdded    w4            1             1
```

FIGURE 8.10
An example of a dataset, pairwise comparison matrix of criteria and alternatives.

```
* A macro to import data;
%macro data;
* To import Criteria Pair;
proc import
      datafile = &_dataC
      out = dataC
      dbms = tab
      replace;
      getnames = no;
run;

* To import Alternative Pair;
 proc import
datafile = &_dataA
      out = dataA
      dbms = tab
      replace;
      getnames = no;
 run;

 data dataA(rename = (VAR1 = Criteria VAR2 = Alternative));
    set dataA;
 run;
%mend data;
```

8.1.6 ORAHP: Model-Building Macro (%model)

This part of ORAHP calls two more macros (%consist and %priority) to solve the AHP model. Two parameters need to be set before calling the model macro (%macro):

_nCriteria: Identities the number of criteria

_nAlternative: Identifies the number of alternatives

The SAS macro for model-building is as follows:

```
* A macro for AHP calculation;
%macro model;

data _NULL_;
array NCarray(&_nCriteria) $ NC1-NC&_nCriteria;
array NAarray(&_nAlternative) $ NA1-NA&_nAlternative;

array PCarray(&_nCriteria,&_nCriteria);
array PAarray(&_nCriteria,&_nAlternative,&_nAlternative);

array TempPAarray(&_nAlternative,&_nAlternative);
array TempAeragePAarray(&_nAlternative);

array AveragePAarray(&_nCriteria,&_nAlternative);
array AveragePCarray(&_nCriteria);

array DividePCarray(&_nCriteria,&_nCriteria);
array SumPCarray(&_nCriteria);

array DividePAarray(&_nAlternative,&_nAlternative);
array SumPAarray(&_nAlternative);

array RankAarray(&_nAlternative);
array LambdaCArray(&_nCriteria);
array LambdaAArray(&_nAlternative);

array RIr(10) RIr1-RIr10;
RIr3 = 0.58;
RIr4 = 0.90;
RIr5 = 1.12;
RIr6 = 1.24;
RIr7 = 1.32;
RIr8 = 1.41;

do i = 1 to &_nCriteria;
      Link ReaddataC;
      NCarray(i) = VAR1;
      %do j = 2 %to &_nCriteria + 1;
            PCarray(i,&j-1) = VAR&j;
      %end;
end;
```

```
do i = 1 to &_nCriteria;
      do j = 1 to &_nAlternative;
            Link ReaddataA;
            NAarray(j) = Alternative;
            %do k = 3 %to &_nAlternative + 2;
                  PAarray(i,j,&k-2) = VAR&k;
            %end;
      end;
end;

do i = 1 to &_nCriteria;
      do j = 1 to &_nCriteria;
            if PCarray(i,j) = . then PCarray(i,j) = 1/
PCarray(j,i);
      end;
end;

do i = 1 to &_nAlternative;
      RankAarray(i) = 0;
end;

do i = 1 to &_nCriteria;
      do j = 1 to &_nAlternative;
            do k = 1 to &_nAlternative;
                  if PAarray(i,j,k) = . then PAarray(i,j,k)
= 1/PAarray(i,k,j);
            end;
      end;
end;

%Priority (PCarray,&_nCriteria,AveragePCarray,DividePCarray
,SumPCarray);
%Consist (&_nCriteria,LambdaCArray,AveragePCarray,PCarray,
Lambda, CI, CR, RI,RIr);
%report1 ("Criteria", &_nCriteria,NCarray,PCarray,AveragePC
array,Lambda, CI, CR, RI);

do m = 1 to &_nCriteria;
      do j = 1 to &_nAlternative;
            do k = 1 to &_nAlternative;
                  TempPAarray(j,k) = PAarray(m,j,k);
            end;
      end;
      %Priority
      (TempPAarray,&_nAlternative,TempAeragePAarray,DivideP
Aarray,SumPAarray);
      do k = 1 to &_nAlternative;
            AveragePAarray(m,k) = TempAeragePAarray(k);
      end;
      %Consist
      (&_nAlternative,LambdaAArray,TempAeragePAarray,TempP
Aarray,Lambda, CI, CR, RI,RIr);
```

```
        %report1 (NCarray (m),&_nAlternative,NAarray,
TempPAarray,TempAeragePAarray,Lambda, CI, CR, RI);
end;

%report2
(&_nAlternative,&_nCriteria,RankAarray,AveragePAarray,Avera
gePCarray);

ReaddataC: set dataC; return;
ReaddataA: set dataA; return;
run;
%mend model;
```

The following macro calculates the consistency of AHP rate:

```
*A macro for calculating AHP measures including consistency;
%macro Consist(nDim, LamArray, AverageArray,PrArray,
Lambda, CI, CR, RI,RIr);

do i1 = 1 to &nDim;
    &LamArray(i1) = 0;
    do j1 = 1 to &nDim;
        &LamArray(i1) = &LamArray(i1) + &AverageArray(j1)
*&PrArray(i1,j1);
    end;
end;

&Lambda = 0;
do i1 = 1 to &nDim;
    &Lambda = &Lambda + &LamArray(i1)/&AverageArray(i1);
end;

&Lambda = &Lambda/&nDim;
&CI = (&Lambda-&nDim)/(&nDim-1);
RI = &RIr(&nDim);
&CR = &CI/RI;
%mend Consist;
```

The following macro calculates the priority of each alternative:

```
* A macro for calculating average priorities;
%macro Priority
(MyMatrix,myDimension,AverageMyMatrix,DivideMyMatrix,SumMy
Matrix);

do j = 1 to &myDimension;
    &SumMyMatrix(j) = 0;
    do i = 1 to &myDimension;
        &SumMyMatrix(j) = &SumMyMatrix(j) +
&MyMatrix(i,j);
    end;
end;
```

```
do i = 1 to &myDimension;
      do j = 1 to &myDimension;
        &DivideMyMatrix(i,j) = &MyMatrix(i,j)/&SumMyMatrix(j);
      end;
end;

do i = 1 to &myDimension;
      &AverageMyMatrix(i) = 0;
      do j = 1 to &myDimension;
            &AverageMyMatrix(i) = &AverageMyMatrix(i) +
&DivideMyMatrix(i,j);
      end;
      &AverageMyMatrix(i) = &AverageMyMatrix(i)/&myDimension;
end;
%mend Priority;
```

8.1.7 ORAHP: Report-Writing Macro (%report1 and %report2)

The outputs from ORAHP include two reports. Report 1 contains all the information, including comparison matrices, priorities, λ_{max}, CI, RI, and CR for each criterion and alternative. Report 2 contains the AHP priority ranking for each alternative.

For report, there are two macros (%report1 and %report2), and one parameter needs to be set before calling these macros:

_title: Gives a title in the output of the SAS

```
* A macro for writing priority matrix;
%macro report1 (Name,nDim,MyName,MyMatrix,AverageMyMatrix,
Lambda, CI, CR, RI);
File Print;
title & Title;
put // " " &Name @@;
do j1 = 1 to &nDim;
    c = 9*j1 + 3;
    put @c &MyName(j1) @@;
end;

c = c + 9; put @c "Priori" @@;
c = c + 9; put @c "LamMax" @@;
c = c + 9; put @c "CI" @@;
c = c + 9; put @c "RI" @@;
c = c + 9; put @c "CR";

do j1 = 1 to &nDim + 5;
    c = 9*j1 + 1; put @c " -------- " @@;
end;
```

```
do i1 = 1 to &nDim;
    put / " " &MyName(i1) @@;
    do j1 = 1 to &nDim;
        c = 9*j1 + 1; put @c &MyMatrix(i1,j1) 8.4 @@;
        end;
  c = c + 10; put @c &AverageMyMatrix(i1) 8.4 @@;
end;

put;
do j1 = 1 to &nDim + 5;
    c = 9*j1 + 1; put @c " -------- " @@;
end;

put; c = 9*&nDim + 10;put @c "Total =  =  = > " @@;

c = &nDim*9 + 20; put @c Lambda 8.4 @@;
c = c + 9; put @c CI 8.4 @@;
c = c + 9; put @c RI 8.4 @@;
c = c + 9; put @c CR 8.4 @@;
%mend report1;
```

```
* A macro for writing AHP results;
%macro report2 (nDimA,nDimC,RankAarray,AveragePAarray,Avera
gePCarray);
do j = 1 to &nDimA;
    RankAarray(j) = 0;
    do i = 1 to &nDimC;
        RankAarray(j) = RankAarray(j) + AveragePAarray(i,j)
*AveragePCarray(i);
    end;
end;

put // @10 "--------------------------" ;
put @10 "AHP Result ";
put @10 &_Title;
put @10 "--------------------------" ;
put @10 "Alternative" @25 "Priority" ;
put @10 "-----------" @25 "---------" ;

do j1 = 1 to &nDimA;
    put @10 NAarray(j1) @@;
    put @25 RankAarray(j1) 8.4;
end;
put @10 "--------------------------" ;
%mend report2;
```

8.1.8 ORAHP: Macro (%orahp)

To make the system as user friendly as possible, the %orahp macro combines the data-handling, model-building, and report-writing codes.

```
* A SAS macro for an AHP (analytic hierarchy process)
problem;
%macro orahp;
 %data;
 %model;
%mend orahp;
```

This code, the %orahp macro is used to manage all the codes explained earlier, including data-handling, model-building, and report-writing. To get the results, the user needs to set up the parameters and run only one statement:

```
%ORAHP;
```

8.1.9 Instructions for Using ORAHP Macro

This section presents SAS code for the earlier example of the warehouse evaluation problem with five criteria and four alternatives using the AHP, as shown in Figure 8.1. The data are saved in files "data8_1_C.txt" and "data8_1_A.txt."

The user needs to set the parameters as required and run the following code:

```
%let _Title = 'Example 8.1. A SAS macro for analytic
hierarchy process (AHP)';
%let _dataC = 'c:/sasor/data8_1_C.txt';
%let _dataA = 'c:/sasor/data8_1_A.txt';
%let _nCriteria = 5;
%let _nAlternative = 4;
options linesize = 100;
%orahp;
```

This code determines the results based on the specified parameters.

8.1.10 Sample Results from ORAHP Macro: Output from SAS

The results from running this code are presented in Figures 8.11 and 8.12. At the top of Figure 8.11, the priorities of five criteria—total lead time, reliability of order fulfillment, quality, flexibility of capacity, and value-added services—are shown. In addition, Figure 8.11 also shows the priorities of four warehouses with respect to each of the five criteria. Because the consistency ratios are all below the maximum 0.10 level, the judgments are consistent and acceptable. The overall priorities of the four warehouses are shown in Figure 8.12. According to this figure, warehouse 1 has the best overall performance because it scores the highest weighting ($wp_1 = 0.4454$), followed by warehouse 3 ($wp_3 = 0.2374$), warehouse 2 ($wp_2 = 0.1827$), and

Output - (Untitled)

Criteria	Lead	Reliabil	Quality	Flexibil	ValueAdd	Priori	LamMax	CI	RI	CR
Lead	1.0000	1.0000	2.0000	3.0000	4.0000	0.3131				
Reliabi	1.0000	1.0000	2.0000	3.0000	4.0000	0.3131				
Quality	0.5000	0.5000	1.0000	3.0000	4.0000	0.2124				
Flexibi	0.3333	0.3333	0.3333	1.0000	1.0000	0.0877				
ValueAd	0.2500	0.2500	0.2500	1.0000	1.0000	0.0739				
					Total ===>		5.0881	0.0220	1.1200	0.0197

Lead	W1	W2	W3	W4	Priori	LamMax	CI	RI	CR
W1	1.0000	3.0000	3.0000	5.0000	0.5136				
W2	0.3333	1.0000	1.0000	4.0000	0.2161				
W3	0.3333	1.0000	1.0000	3.0000	0.1968				
W4	0.2000	0.2500	0.3333	1.0000	0.0735				
				Total ===>		4.0761	0.0254	0.9000	0.0282

Reliabil	W1	W2	W3	W4	Priori	LamMax	CI	RI	CR
W1	1.0000	3.0000	2.0000	4.0000	0.4658				
W2	0.3333	1.0000	0.5000	2.0000	0.1611				
W3	0.5000	2.0000	1.0000	3.0000	0.2771				
W4	0.2500	0.5000	0.3333	1.0000	0.0960				
				Total ===>		4.0310	0.0103	0.9000	0.0115

Quality	W1	W2	W3	W4	Priori	LamMax	CI	RI	CR
W1	1.0000	3.0000	2.0000	3.0000	0.4547				
W2	0.3333	1.0000	0.5000	1.0000	0.1411				
W3	0.5000	2.0000	1.0000	2.0000	0.2630				
W4	0.3333	1.0000	0.5000	1.0000	0.1411				
				Total ===>		4.0104	0.0035	0.9000	0.0038

Flexibil	W1	W2	W3	W4	Priori	LamMax	CI	RI	CR
W1	1.0000	0.3333	0.5000	0.2000	0.0867				
W2	3.0000	1.0000	3.0000	0.5000	0.2978				
W3	2.0000	0.3333	1.0000	0.3333	0.1425				
W4	5.0000	2.0000	3.0000	1.0000	0.4730				
				Total ===>		4.0651	0.0217	0.9000	0.0241

ValueAdd	W1	W2	W3	W4	Priori	LamMax	CI	RI	CR
W1	1.0000	4.0000	2.0000	3.0000	0.4687				
W2	0.2500	1.0000	0.3333	1.0000	0.1152				
W3	0.5000	3.0000	1.0000	2.0000	0.2800				
W4	0.3333	1.0000	0.5000	1.0000	0.1361				

FIGURE 8.11
Results of %ORAHP.

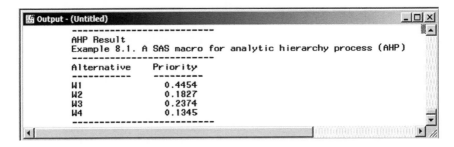

Output - (Untitled)

```
--------------------------------
AHP Result
Example 8.1. A SAS macro for analytic hierarchy process (AHP)
--------------------------------
Alternative      Priority
-----------      --------
   W1             0.4454
   W2             0.1827
   W3             0.2374
   W4             0.1345
--------------------------------
```

FIGURE 8.12
Results of % ORAHP.

warehouse 4 ($wp_4 = 0.1345$). The AHP priorities are used to determine the priority level of the AHP priority constraints in the goal-programming model (see Section 8.2).

8.1.11 Exercise

Use the codes developed in this chapter and solve the AHP problem presented in Figures 8.13 to 8.18.
Solution:

- Create the data in a text file (see "data8_1_A_exercise.txt" and "data8_1_C_exercise.txt").
- Run the following code (see program "sasor_8_1_exercise.sas").

```
* SAS macro for Analytical Hierarchy Process: solution to
exercise 8.1.;
%let _title = 'Analytical Hierarchy Process, solution to
exercise 8.1';
%let _dataC = 'c:/sasor/data8_1_C_ exercise.txt';
%let _dataA = 'c:/sasor/data8_1_A_ exercise.txt';
%let _nCriteria = 4;
%let _nAlternative = 3;
options linesize = 100;
%orahp;
```

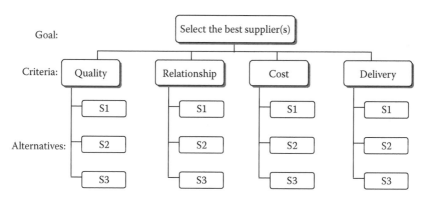

FIGURE 8.13
The hierarchy of supplier prioritization. S1, supplier 1; S2, supplier 2; S3, supplier 3.

$$A_1 = \begin{bmatrix} 1 & 3 & 2 & 2 \\ \dfrac{1}{3} & 1 & \dfrac{1}{4} & \dfrac{1}{4} \\ \dfrac{1}{2} & 4 & 1 & \dfrac{1}{2} \\ \dfrac{1}{2} & 4 & 2 & 1 \end{bmatrix}$$

FIGURE 8.14
Pairwise comparison matrix for criteria with respect to goal.

$$A_2 = \begin{bmatrix} 1 & \dfrac{1}{4} & \dfrac{1}{5} \\ 4 & 1 & \dfrac{1}{3} \\ 5 & 3 & 1 \end{bmatrix}$$

FIGURE 8.15
Pairwise comparison matrix for suppliers
with respect to quality.

$$A_4 = \begin{bmatrix} 1 & 3 & 9 \\ \dfrac{1}{3} & 1 & 7 \\ \dfrac{1}{9} & \dfrac{1}{7} & 1 \end{bmatrix}$$

FIGURE 8.17
Pairwise comparison matrix for suppliers
with respect to cost.

$$A_3 = \begin{bmatrix} 1 & \dfrac{1}{5} & \dfrac{1}{7} \\ 5 & 1 & \dfrac{1}{4} \\ 7 & 4 & 1 \end{bmatrix}$$

FIGURE 8.16
Pairwise comparison matrix for suppliers
with respect to relationship.

$$A_5 = \begin{bmatrix} 1 & \dfrac{1}{4} & 5 \\ 4 & 1 & 8 \\ \dfrac{1}{5} & \dfrac{1}{8} & 1 \end{bmatrix}$$

FIGURE 8.18
Pairwise comparison matrix for suppliers
with respect to delivery.

The solution shown in Figures 8.19 and 8.20 is given by SAS.

8.2 Multiple-Criteria Logistics Distribution Problem: Phase 2

8.2.1 Example of Multiple-Criteria Logistics Distribution Problem

In Section 8.1, we applied the AHP to evaluate the relative importance weightings of four alternative warehouses with respect to five qualitative-based evaluating criteria. In this section, a GP model is formulated to select the optimal set of warehouses while considering the AHP priorities of warehouses and the quantitative-based limitations of resources.

```
Output - (Untitled)                                                      _|□|×|

Criteria  Quality  Relation Cost    Delivery Priori  LamMax    CI       RI       CR
          -------- -------- -------- -------- ------- --------  -------- -------- --------
Quality   1.0000   1.0000   3.0000   2.0000   0.3666
Relatio   1.0000   1.0000   0.2500   0.2500   0.1399
Cost      0.3333   4.0000   1.0000   0.5000   0.2027
Deliver   0.5000   4.0000   2.0000   1.0000   0.2908
          -------- -------- -------- -------- -------           -------- -------- --------
                                     Total ===> 4.7897  0.2632   0.9000   0.2925

Quality   S1       S2       S3       Priori   LamMax    CI       RI       CR
          -------- -------- -------- -------- ------- --------  -------- -------- --------
S1        1.0000   0.2500   0.2000   0.0964
S2        4.0000   1.0000   0.3333   0.2842
S3        5.0000   3.0000   1.0000   0.6194
          -------- -------- -------- -------- -------           -------- -------- --------
                                     Total ===> 3.0867  0.0433   0.5800   0.0747

Relation  S1       S2       S3       Priori   LamMax    CI       RI       CR
          -------- -------- -------- -------- ------- --------  -------- -------- --------
S1        1.0000   0.2000   0.1429   0.0726
S2        5.0000   1.0000   0.2500   0.2521
S3        7.0000   4.0000   1.0000   0.6752
          -------- -------- -------- -------- -------           -------- -------- --------
                                     Total ===> 3.1263  0.0631   0.5800   0.1089

Cost      S1       S2       S3       Priori   LamMax    CI       RI       CR
          -------- -------- -------- -------- ------- --------  -------- -------- --------
S1        1.0000   3.0000   9.0000   0.6486
S2        0.3333   1.0000   7.0000   0.2946
S3        0.1111   0.1429   1.0000   0.0567
          -------- -------- -------- -------- -------           -------- -------- --------
                                     Total ===> 3.0813  0.0406   0.5800   0.0701

Delivery  S1       S2       S3       Priori   LamMax    CI       RI       CR
          -------- -------- -------- -------- ------- --------  -------- -------- --------
S1        1.0000   0.2500   5.0000   0.2438
S2        4.0000   1.0000   8.0000   0.6893
S3        0.2000   0.1250   1.0000   0.0669
          -------- -------- -------- -------- -------  -------- -------- --------
                                     Total ===> 3.0956  0.0478   0.5800   0.0824
```

FIGURE 8.19
Results of %ORAHP.

```
Output - (Untitled)                                                      _|□|×|

--------------------------
AHP Result
Analytical Hierarchy Process, solution to exercise 8.1
--------------------------
Alternative    Priority
-----------    --------
S1             0.2479
S2             0.3996
S3             0.3525
--------------------------
```

FIGURE 8.20
Results of % ORAHP.

Consider a typical logistics distribution network that consists of m warehouses denoted as $i = \{1, 2, ..., m\}$ and n customers denoted as $j = \{1, 2, ..., n\}$. Each warehouse has;

- A maximum throughput, Q_i
- A minimum throughput, q_i
- A fixed cost, fc_i
- A unit inventory holding cost, hc_i

Each customer has a unique order volume, D_j. When warehouse i is assigned to serve customer j, it costs dc_{ij} dollars per unit for delivery. If the total amount of products assigned to warehouse i (i.e., $\sum_{j=1}^{n} x_{ij},\ \forall i$) is less than q_i, this is regarded as an impractical allocation because it is not cost-effective to set up a warehouse for processing only a few orders. To avoid low effectiveness of warehouse utilization, penalty cost, pc_i, is considered in the model, which is incurred if $0 < \sum_{j=1}^{n} x_{ij} < q_i$. The problem here is to determine an *optimal distribution network*, which refers to the most cost-efficient allocation of orders to the available warehouses (Ho and Emrouznejad, 2009).

Before formulating the GP model for the design of a logistics distribution network, data on coefficients and right-side value should be collected. The necessary resource data are presented in Tables 8.3 and 8.4.

In the model, there are four decision variables:

x_{ij} = amount of products delivered from warehouse i to customer j

$$u_i = \begin{cases} 1 & \text{if total allocation of products to warehouse } i \text{ is less than } q_i \\ 0 & \text{otherwise} \end{cases}$$

TABLE 8.3

Resource Data for the GP Model

| Warehouse, i | Unit Delivery Cost ($), dc_{ij} | | | | | | |
| | Customer, j | | | | | | |
	1	2	3	4	5	6	7
1	1	1	2	4	4	3	6
2	2	6	9	3	7	8	4
3	8	4	3	6	3	2	4
4	8	8	9	3	5	7	2
Amount required by customer j, D_j	12,000	9000	10,000	8000	6000	11,000	7000

TABLE 8.4

Resource Data for the GP Model (Continued)

Warehouse, i	Maximum Throughput of Warehouse i Q_i	Minimum Throughput of Warehouse i q_i	Unit Holding Cost ($) hc_i	Fixed Cost ($) fc_i	Penalty Cost ($) pc_i
1	30,000	6000	5	30,000	7500
2	26,000	5200	3	25,000	6250
3	22,000	4400	3	20,000	5000
4	18,000	3600	2	15,000	3750

Targeted total cost, TC = $425,000; Arbitrary large number, M = 100,000.

$$
v_i = \begin{cases} 1 & \text{if warehouse } i \text{ is selected} \\ 0 & \text{otherwise} \end{cases}
$$

$$
w_i = \begin{cases} 1 & \text{if both } u_i \text{ and } v_i \text{ equal to one} \\ 1 & \text{otherwise} \end{cases}
$$

In the model, there are three types of constraints: system constraints, resource constraints, and AHP priority constraints. *System constraints* are ordinary linear programming constraints in which there is no deviation variable. This type of constraint cannot be violated and thus is called a hard constraint. *Resource constraints* are goal equations, or soft constraints, in which there are deviation variables. *AHP priority constraints* are akin to resource constraints. This type of constraint has deviation variables, the priority levels of which are dependent on the overall AHP priority ranking.

Model 8.2.1 Standard multiple-criteria logistics distribution model
System constraints

$$
\sum_{i=1}^{m} v_i \leq m \tag{8.2.1}
$$

$$
\sum_{j=1}^{n} x_{ij} + M u_i \geq q_i \quad \text{for } i = 1, 2, \ldots, m \tag{8.2.2}
$$

$$
\sum_{j=1}^{n} x_{ij} - M v_i \leq 0 \quad \text{for } i = 1, 2, \ldots, m \tag{8.2.3}
$$

$$
w_i - u_i - v_i = -1 \quad \text{for } i = 1, 2, \ldots, m \tag{8.2.4}
$$

Resource constraints
Priority 1 (P_1):

$$
\sum_{j=1}^{n} x_{ij} - d_i^+ + d_i^- = Q_i \quad \text{for } i = 1, 2, \ldots, m \tag{8.2.5}
$$

$$
\sum_{i=1}^{m} x_{ij} - d_{j+m}^+ + d_{j+m}^+ = D_j \quad \text{for } j = 1, 2, \ldots, n \tag{8.2.6}
$$

Priority 2 (P_2):

$$\sum_{i=1}^{m}\sum_{j=1}^{n}\left(hc_i+dc_{ij}\right)x_{ij}+\sum_{i=1}^{m}fc_iv_i-d^+_{m+n+1}+d^-_{m+n+1}=TC \qquad (8.2.7)$$

Priority 3 (P_3):

$$\sum_{i=1}^{m}pc_iw_i-d^+_{m+n+2}+d^-_{m+n+2}=0 \qquad (8.2.8)$$

AHP priority constraints
  Priorities 4 to $4+m-1$ (P_4 to P_{4+m-1}):

$$v_i-d^+_{m+n+2+i}+d^-_{m+n+2+i}=1 \quad \text{for } i=1,2,\dots,m \qquad (8.2.9)$$

Objective function

$$\text{Minimize } z=\sum_{k=1}^{3}P_k\left(d^+_r+d^-_r\right)+\sum_{k=4}^{m+3}P_k\left(d^+_s+d^-_s\right)$$

$$\text{for } r=1,2,\dots,m+n+2$$

$$\text{for } s=m+n+3,\,m+n+4,\dots,2m+n+2 \qquad (8.2.10)$$

Model 8.2.1 is referred to as the *GP model*. Constraint set 8.2.1 ensures that the number of warehouses selected must be equal to or less than the number of warehouses available. Constraint set 8.2.2 determines which warehouses have an allocation of products that are less than the minimum warehouse throughput. Constraint set 8.2.3 determines which warehouses are selected. Constraint set 8.2.4 determines which warehouses incur penalty costs. Constraint set 8.2.5 allocates products to warehouses while keeping the amount less than the maximum warehouse throughput. Constraint set 8.2.6 allocates products to warehouses while keeping the amount equal to that demanded by the customers. Constraint set 8.2.7 ensures that the total cost, including the inventory holding cost, delivery cost, and fixed cost associated with warehouse selection, does not exceed the targeted amount. Constraint set 8.2.8 ensures that products are not allocated to warehouses incurring penalty costs. Constraint set 8.2.9 selects warehouse i. Objective function 8.2.10 minimizes the total deviations from the goals.

According to Section 8.1, warehouse 1 has the best overall performance because it scores the highest weighting ($wp_i=0.4454$), followed by warehouse 3 ($wp_3=0.2374$), warehouse 2 ($wp_2=0.1827$), and warehouse 4 ($wp_4=0.1345$).

The AHP priorities are used to determine the priority level of the AHP priority constraints—that is, constraint set 8.2.9.

Based on the AHP priorities of four alternative warehouses and resource data in Tables 8.3 and 8.4, the GP model for the logistics distribution problem can be formulated as shown in Model 8.2.2.

Model 8.2.2 Example of formulation of multiple-criteria logistics distribution problem

$$\text{Minimize } z = P_1\left[\sum_{k=1}^{4} d_k^+ + \sum_{k=5}^{11}\left(d_k^+ + d_k^-\right)\right] + P_2\left(d_{12}^+\right) + P_3\left(d_{13}^+\right)$$

$$+ P_4\left(d_{14}^+ + d_{14}^-\right) + P_5\left(d_{15}^+ + d_{15}^-\right) + P_6\left(d_{16}^+ + d_{16}^-\right) + P_7\left(d_{17}^+ + d_{17}^-\right) \tag{8.2.11}$$

subject to

$$v_1 + v_2 + v_3 + v_4 \le 4 \tag{8.2.12}$$

$$x_{11} + x_{12} + x_{13} + x_{14} + x_{15} + x_{16} + x_{17} + 100{,}000\, u_1 \ge 6000 \tag{8.2.13}$$

$$x_{21} + x_{22} + x_{23} + x_{24} + x_{25} + x_{26} + x_{27} + 100{,}000\, u_2 \ge 5200 \tag{8.2.14}$$

$$x_{31} + x_{32} + x_{33} + x_{34} + x_{35} + x_{36} + x_{37} + 100{,}000\, u_3 \ge 4400 \tag{8.2.15}$$

$$x_{41} + x_{42} + x_{43} + x_{44} + x_{45} + x_{46} + x_{47} + 100{,}000\, u_4 \ge 3600 \tag{8.2.16}$$

$$x_{11} + x_{12} + x_{13} + x_{14} + x_{15} + x_{16} + x_{17} - 100{,}000\, v_1 \le 0 \tag{8.2.17}$$

$$x_{21} + x_{22} + x_{23} + x_{24} + x_{25} + x_{26} + x_{27} - 100{,}000\, v_2 \le 0 \tag{8.2.18}$$

$$x_{31} + x_{32} + x_{33} + x_{34} + x_{35} + x_{36} + x_{37} - 100{,}000\, v_3 \le 0 \tag{8.2.19}$$

$$x_{41} + x_{42} + x_{43} + x_{44} + x_{45} + x_{46} + x_{47} - 100{,}000\, v_4 \le 0 \tag{8.2.20}$$

$$w_1 - u_1 - v_1 = -1 \tag{8.2.21}$$

$$w_2 - u_2 - v_2 = -1 \tag{8.2.22}$$

$$w_3 - u_3 - v_3 = -1 \tag{8.2.23}$$

$$w_4 - u_4 - v_4 = -1 \tag{8.2.24}$$

$$P_1 : x_{11} + x_{12} + x_{13} + x_{14} + x_{15} + x_{16} + x_{17} - d_1^+ + d_1^- = 30,000 \qquad (8.2.25)$$

$$x_{21} + x_{22} + x_{23} + x_{24} + x_{25} + x_{26} + x_{27} - d_2^+ + d_2^- = 26,000 \qquad (8.2.26)$$

$$x_{31} + x_{32} + x_{33} + x_{34} + x_{35} + x_{36} + x_{37} - d_3^+ + d_3^- = 22,000 \qquad (8.2.27)$$

$$x_{41} + x_{42} + x_{43} + x_{44} + x_{45} + x_{46} + x_{47} - d_4^+ + d_4^- = 18,000 \qquad (8.2.28)$$

$$x_{11} + x_{21} + x_{31} + x_{41} - d_5^+ + d_5^- = 12,000 \qquad (8.2.29)$$

$$x_{12} + x_{22} + x_{32} + x_{42} - d_6^+ + d_6^- = 9000 \qquad (8.2.30)$$

$$x_{13} + x_{23} + x_{33} + x_{43} - d_7^+ + d_7^- = 10,000 \qquad (8.2.31)$$

$$x_{14} + x_{24} + x_{34} + x_{44} - d_8^+ + d_8^- = 8000 \qquad (8.2.32)$$

$$x_{15} + x_{25} + x_{35} + x_{45} - d_9^+ + d_9^- = 6000 \qquad (8.2.33)$$

$$x_{16} + x_{26} + x_{36} + x_{46} - d_{10}^+ + d_{10}^- = 11,000 \qquad (8.2.34)$$

$$x_{17} + x_{27} + x_{37} + x_{47} - d_{11}^+ + d_{11}^- = 7000 \qquad (8.2.35)$$

$$P_2 : 6x_{11} + 6x_{12} + 7x_{13} + 9x_{14} + 9x_{15} + 8x_{16} + 11x_{17} + 30,000\, v_1$$

$$+5x_{21} + 9x_{22} + 12x_{23} + 6x_{24} + 10x_{25} + 11x_{26} + 7x_{27} + 25,000\, v_2$$

$$+11x_{31} + 7x_{32} + 6x_{33} + 9x_{34} + 6x_{35} + 5x_{36} + 7x_{37} + 20,000\, v_3$$

$$+10x_{41} + 10x_{42} + 11x_{43} + 5x_{44} + 7x_{45} + 9x_{46}$$

$$+4x_{47} + 15,000v_4 - d_{12}^+ + d_{12}^- = 42,5000 \qquad (8.2.36)$$

$$P_3 : 7500w_1 + 6250w_2 + 5000w_3 + 3750w_4 - d_{13}^+ + d_{13}^- = 0 \qquad (8.2.37)$$

$$P_4 : v_1 - d_{14}^+ + d_{14}^- = 1 \qquad (8.2.38)$$

$$P_5 : v_3 - d_{15}^+ + d_{15}^- = 1 \qquad (8.2.39)$$

$$P_6 : v_2 - d_{16}^+ + d_{16}^- = 1 \qquad (8.2.40)$$

$$P_7 : v_4 - d_{17}^+ + d_{17}^- = 1 \qquad (8.2.41)$$

Model 8.2.2 has 13 system constraints (constraint sets 8.2.12 to 8.2.24), 13 resource constraints (constraint sets 8.2.25 to 8.2.37), and four AHP priority constraints (constraint sets 8.2.38 to 8.2.41). Goals with priority level P_1 are the most important, followed by those with priority level P_2, and so on (i.e., $P_1 > P_2 > \ldots > P_7$). Those with a higher priority level are considered first. Once they have been satisfied that there can be no further improvement, the next most important goals are then considered. In P_4 to P_7, the goal level of warehouse i is dependent on its AHP priority ranking (i.e., wp_i). Because warehouse 1 has the best overall performance ($wp_1 = 0.4454$), it is in P_4, followed by warehouse 3 ($wp_3 = 0.2374$), warehouse 2 ($wp_2 = 0.1827$), and warehouse 4 ($wp_4 = 0.1345$).

8.2.2 ORMCDM: SAS Code for Multiple-Criteria Logistics Distribution Problem

ORMCDM is a macro that solves multiple-criteria decision-making problems formulated in goal programming. In the example, the objective is to determine an *optimal distribution network,* which refers to the allocation of orders to the warehouses that will solve the problem to optimality. The procedure used for the multiple-criteria logistics distribution problem is PROC OPTMODEL.

Figure 8.21 illustrates the data flow in ORMCDM. It shows:

- The cost matrix and demand values that are required for ORMCDM
- The macros (%data, %model, and %report),
- The results datasets that are available for print or can be used for further analysis

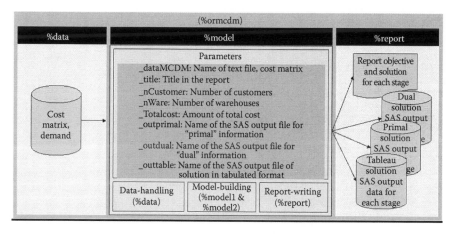

FIGURE 8.21
Data flow in ORMCDM.

In the rest of this section, the procedure for implementing a multiple-criteria decision-making problem (ORMCDM) in SAS, together with an example, is explained. ORMCDM runs four macros: data-handling (%data), model-building (%model), and report-writing (%report).

8.2.3 ORMCDM: Data-Handling Macro (%data)

This part of ORMCDM processes the data into a format suitable for use in PROC OPTMODEL (see program "sasor_8_2.sas"). ORMCDM requires one dataset that contains the names of the warehouses and customers and the cost matrix. The dataset should be a .txt file that is saved as "text tab delimited." The warehouses' and customers' names must start with a letter and may contain up to 50 characters. The variable names must be listed in the first row of the data file. The warehouses' names should be listed in the first column. It should be noted that the demand should be in the last row of the dataset. Other information, including Q_i, q_i, hc_i, fc_i, and pc_i, should also be a part of the data set, but the order of the columns is not important. An example data file is shown in Figure 8.22.

One parameter needs to be set before calling the data macro:

_data: Indicates the name and location of the data file (a text tab delimited file) and contains cost matrix

```
* The data-handling macro;
%macro data;
* Import text tab delimited data file to SAS data file;
  proc import
      datafile = &_dataMCDM
      out = dataC
      dbms = tab
      replace;
      getnames = yes;
  run;
%mend data;
```

War eh	Cust1	Cust2	Cust3	Cust4	Cust5	Cust6	Cust7	QQi	qi	hci	fci	pci
w1	1	1	2	4	4	3	6	30000	6000	5	30000	7500
w2	2	6	9	3	7	8	4	26000	5200	3	25000	6250
w3	8	4	3	6	3	2	4	22000	4400	3	20000	5000
w4	8	8	9	3	5	7	2	18000	3600	2	15000	3750
demand	12000	9000	10000	8000	6000	11000	7000					

FIGURE 8.22
An example of a dataset, cost, and other information for warehouses and customers.

8.2.4 ORMCDM: Model-Building Macro (%model)

This part of ORMCDM calls PROC OPTMODEL to solve the model. There is one SAS macro for model-building. Macro %model creates a suitable data-set that can be used for PROC OPTMODEL. Four parameters need to be set before calling the model macro:

_nWareh: Defines the number of warehouses

_nCustomer: Defines the number of customers

_TotalCost: Defines the total cost

_myupperbnd: Defines the upper bound value for integer variables

```
* The model-building macro;
%macro model (_stage);
* Starting OPTMODEL Procedure;
proc optmodel;

* Define sets;
set WAREHOUSE = 1..&_nWareh;
set CUSTOMER = 1..&_nCustomer;
set WAREHOUSE_CUSTOMER_TWO = 1..&_nCustomer + 2*&_nWareh + 2;
set Priority = 1..&_nWareh + 3;
set stage ;
stage = {&_stage};

* Define variables;
var X{WAREHOUSE, CUSTOMER} integer > = 0;
var u {WAREHOUSE} binary ;
var v {WAREHOUSE} binary ;
var w {WAREHOUSE} binary ;
var dPlus {WAREHOUSE_CUSTOMER_TWO} > = 0;
var dNeg {WAREHOUSE_CUSTOMER_TWO} > = 0;
var z{Priority};

* Define parameters;
number dc{WAREHOUSE, CUSTOMER};
number demand{CUSTOMER};
number QQ{WAREHOUSE};
number q{WAREHOUSE};
number hc{WAREHOUSE};
number fc{WAREHOUSE};
number pc{WAREHOUSE};
number WarehPriority{WAREHOUSE};
number P{Priority, Priority};

for {i in Priority , j in Priority: i ne j} P[i,j] = 0;
for {i in Priority , j in Priority: i eq j} P[i,j] = 1;
```

```
* Load the unit delivery cost matrix;
read data Datac (where = (Wareh ne "demand"))
into [_N_]
{j in CUSTOMER} < dc[_N_,j] = col('Cust'||j) >;

* Load the demand array;
read data Datac (where = (Wareh eq "demand"))
into
{j in CUSTOMER} < demand[j] = col('Cust'||j) >;

* Load the maximum throughput of warehouse array;
read data Datac (where = (Wareh ne "demand"))
into [_N_]
 QQ[_N_] = col('QQi');

* Load the minimum throughput of warehouse array;
read data Datac (where = (Wareh ne "demand"))
into [_N_]
q[_N_] = col('qi');

* Load the unit holding costarray;
read data Datac (where = (Wareh ne "demand"))
into [_N_]
hc[_N_] = col('hci');

* Load the fixed costarray;
read data Datac (where = (Wareh ne "demand"))
into [_N_]
fc[_N_] = col('fci');

* Load the penalty costarray;
read data Datac (where = (Wareh ne "demand"))
into [_N_]
pc[_N_] = col('pci');

* Load the warehouse Priority that is obtained from AHP;
read data Datac (where = (Wareh ne "demand"))
into [_N_]
WarehPriority[_N_] = col('WarehPriority');

* Define system constrains;
con sumV: sum{i in WAREHOUSE} v[i] < = &_nWareh;
con Mu {i in WAREHOUSE}: sum{j in CUSTOMER} x[i,j] + &_
myLarge * u[i] < =q[i];
con Mv {i in WAREHOUSE}: sum{j in CUSTOMER} x[i,j] - &_
myLarge * v[i] < =0;
con wuv {i in WAREHOUSE}: w[i] - u[i] - v[i] = -1;

* Define resource constraints;

* Priority 1 (P1);
con P1_1{ s in Priority, i in WAREHOUSE: s > =1}: sum{j in
CUSTOMER} x[i,j] - dPlus[i] + dNeg[i] < =QQ[i];
```

```
* Priority 2 (P2);
con P2{s in Priority : s > = 2}: sum{i in WAREHOUSE,j in
CUSTOMER} (hc[i] + dc[i,j])*x[i,j] + sum{i in WAREHOUSE}
fc[i]*v[i] - dPlus[&_nCustomer + &_nWareh + 1] + dNeg[&_
nCustomer + &_nWareh + 1] < = &_TotalCost;

* Priority 3 (P3);
con P3{s in Priority : s > = 3}: sum{i in WAREHOUSE}
pc[i]*w[i] - dPlus[&_nCustomer + &_nWareh + 2] + dNeg[&_
nCustomer + &_nWareh + 2] = 0;

* Define AHP (P4) priority constraints;
con P4{s in Priority, i in WAREHOUSE: s > = i + 3}:
v[WarehPriority[i]]- dPlus[&_nCustomer + &_nWareh + i + 2]
+ dNeg[&_nCustomer + &_nWareh + i + 2] = 1;

con z [1] = (sum {k in 1..&_nWareh} dPlus[k] + sum {k in
&_nWareh + 1..&_nWareh + &_nCustomer} (dPlus[k] + dNeg[k]));
con z [2] = (dPlus[&_nWareh + &_nCustomer + 1]);
con z [3] = (dPlus[&_nWareh + &_nCustomer + 2]);
con z [4] = (dPlus[&_nWareh + &_nCustomer + 3] + dNeg[&_
nWareh + &_nCustomer + 3]);
con z [5] = (dPlus[&_nWareh + &_nCustomer + 4] + dNeg[&_
nWareh + &_nCustomer + 4]);
con z [6] = (dPlus[&_nWareh + &_nCustomer + 5] + dNeg[&_
nWareh + &_nCustomer + 5]);
con z [7] = (dPlus[&_nWareh + &_nCustomer + 6] + dNeg[&_
nWareh + &_nCustomer + 6]);
* Define objective function;

* Objective function;
min obj = sum {s1 in Priority, s in stage} P[s1 , s ]*
z[s1];

solve with MILP; %put &_OROPTMODEL_; expand;

create data _optimout
from [WAREHOUSE CUSTOMER]
 = {i in WAREHOUSE, j in CUSTOMER}
amount = X;

quit;

%mend model;
```

Macro %model defines seven optimization models that could be used to solve the multiple-criteria decision-making problem in seven stages (or priority levels). In each stage, a linear program is created, then PROC OPTMODEL solves the model and the results are saved in a data set for future analysis. One parameter needs to be set before calling this macro:

_myLarge: Defines an optional large number

The parameter _myLarge is an optional parameter that is needed for linear programming, and in this case, it has been set to 1,000,000.

8.2.5 ORMCDM: Report-Writing Macro (%report)

The outputs from ORMCDM include one report for each stage. This report contains all the information regarding the summary and solution of each stage of the multiple-criteria decision-making problem—in this case, seven stages. All this information is also saved in SAS datasets. The user needs to define appropriate names for each of these datasets before calling %ormcdm macro:

_outprimal: Identifies the name of the SAS output file for primal solution

_outdual: Identifies the name of the SAS output file for dual solution

_outtable: Identifies the name of the SAS output file for tabulated solution

Another parameter needs to be set before calling this macro:

_title: Gives a title in the output of the SAS

```
* The report-writing macro;
%macro report (_stage);
 * Report the results in a tabulated form;
title &_title '(P = ' &_stage ')';
 proc tabulate data = _optimout;
 title &_title;
 class WAREHOUSE CUSTOMER ;
 var amount;
 table WAREHOUSE = "WAREHOUSE",
     CUSTOMER*amount*sum
     / BOX = 'x[WAREHOUSE& CUSTOMER]' ;
 run;
%mend report;
```

8.2.6 ORMCDM: Macro (%ormcdm)

To make the system as user friendly as possible, the %ormcdm macro combines the data-handling, model-building, and report-writing codes.

```
* A SAS macro for multiple criteria decision making
problem;
%macro ormcdm;
 %data;
 %do i = 1 %to &_nWareh + 3;
 %model(&i);
 %report(&i);
 %end;
%mend ormcdm;
```

In this code, the %ormcdm macro is used to manage all the codes explained earlier, including data-handling, model-building, and report-writing. To get the result, user needs to set up the parameters and run only one statement:

```
%ORMCDM;
```

8.2.7 Instructions for Using ORMCDM Macro

This section presents SAS code for the earlier example of the multiple-criteria logistics distribution problem with four warehouses and seven customers as shown in Tables 8.3 and 8.4. The data is saved in file "data8_2.txt."

The user needs to set the parameters as required and run the following code:

```
options nodate;
%let _data = 'c:\sasor\MCDM.txt';
%let _title = 'Example 8.2: An example of MCDM';
%let _dataMCDM = 'c:/sasor/Data8_2.txt';
%let _nWareh = 4;
%let _nCustomer = 7;
%let _TotalCost = 425000;
%let _outprimal = outprimal;
%let _outdual = outdual;
%let _outtable = outtable;
%let _myLarge = 100000;
%ormcdm;
```

This code determines the results based on the specified parameters and the cost matrix saved in the text file and also produces a macro variable (_OROPTMODEL_) at termination. Users can examine the result of this macro variable, examine whether PROC OPTMODEL ran correctly, and examine what error or difficulty it encountered. A summary of information, including the objective value at optimum level and the status of _OROPTMODEL_, can be seen in the log file as shown in Figure 8.23.

8.2.8 Sample Results from ORMCDM Macro: Output from SAS

Some of the results of running this code are presented in Figure 8.24 and Figure 8.25, which summarize the solutions in stage (or priority level) 5 and stage 6, respectively. The solution, shown in Figure 8.23, is feasible because the allocation does not exceed the maximum throughput of warehouses, satisfies the volume requirement of customers, remains within the total cost budget, and does not incur any penalty cost. When stage 6 was found to be unachievable (i.e., $d_{16}^- = 1$) (see Figure 8.24), the optimization process was

NOTE: There were 31 observations read from the dataset WORK.DATAMCDM1. ——— Stage 1
NOTE: The dataset WORK.OUTDUAL 1 has 31 observations and 9 variables.
NOTE: The dataset WORK.OUTPRIMAL 1 has 85 observations and 10 variables.
NOTE: The dataset WORK.OUTTABLE 1 has 33 observations and 89 variables.
NOTE: PROCEDURE LP used (Total process time):
    real time   0.07 seconds
    cpu time   0.04 seconds

STATUS=SUCCESSFUL PHASE=2 OBJECTIVE=0 P_FEAS=YES D_FEAS=YES PHASE1_I
TER=17 PHASE2_I TER=4 PHASE3_I TER=0

NOTE: There were 48 observations read from the dataset WORK.DATAMCDM2. ——— Stage 2
NOTE: The dataset WORK.OUTDUAL2 has 48 observations and 9 variables.
NOTE: The dataset WORK.OUTPRIMAL2 has 85 observations and 10 variables.
NOTE: The dataset WORK.OUTTABLE2 has 51 observations and 89 variables.
NOTE: PROCEDURE LP used (Total process time):
    real time   0.04 seconds
    cpu time   0.03 seconds

STATUS=SUCCESSFUL PHASE=2 OBJECTIVE=0 P_FEAS=YES D_FEAS=YES PHASE1_I
TER=14 PHASE2_I TER=0 PHASE3_I TER=0

NOTE: There were 50 observations read from the dataset WORK.DATAMCDM3. ——— Stage 3
NOTE: The dataset WORK.OUTDUAL3 has 50 observations and 9 variables.
NOTE: The dataset WORK.OUTPRIMAL3 has 85 observations and 10 variables.
NOTE: The dataset WORK.OUTTABLE3 has 52 observations and 89 variables.
NOTE: PROCEDURE LP used (Total process time):
    real time   0.09 seconds
    cpu time   0.04 seconds

STATUS=SUCCESSFUL PHASE=2 OBJECTIVE=0 P_FEAS=YES D_FEAS=YES PHASE1_I
TER=12 PHASE2_I TER=0 PHASE3 I TER=0

NOTE: There were 51 observations read from the dataset WORK.DATAMCDM4. ——— Stage 4
NOTE: The dataset WORK.OUTDUAL4 has 51 observations and 9 variables.
NOTE: The dataset WORK.OUTPRIMAL4 has 85 observations and 10 variables.
NOTE: The dataset WORK.OUTTABLE4 has 53 observations and 89 variables.
NOTE: PROCEDURE LP used (Total process time):
    real time   0.04 seconds
    cpu time   0.03 seconds

STATUS=SUCCESSFUL PHASE=2 OBJECTIVE=0 P_FEAS=YES D_FEAS=YES PHASE1_I
TER=1 PHASE2_I TER=1 PHASE3_I TER=0

NOTE: There were 53 observations read from the dataset WORK.DATAMCDM5. ——— Stage 5
NOTE: The dataset WORK.OUTDUAL5 has 53 observations and 9 variables.
NOTE: The dataset WORK.OUTPRIMAL5 has 85 observations and 10 variables.
NOTE: The dataset WORK.OUTTABLE5 has 55 observations and 89 variables.
NOTE: PROCEDURE LP used (Total process time):
    real time   0.10 seconds
    cpu time   0.01 seconds

STATUS=SUCCESSFUL PHASE=2 OBJECTIVE=0 P_FEAS=YES D_FEAS=YES PHASEI_I
TER=12 PHASE2_I TER=0 PHASE3_I TER=0

FIGURE 8.23
Log for %ORMCDM.

NOTE: There were 55 observations read from the dataset WORK.DATAMCDM6. ———— Stage 6
NOTE: The dataset WORK.OUTDUAL6 has 55 observations and 9 variables.
NOTE: The dataset WORK.OUTPRIMAL6 has 85 observations and 10 variables.
NOTE: The dataset WORK.OUTTABLE6 has 57 observations and 89 variables.
NOTE: PROCEDURE LP used (Total process time):
     real time   0.03 seconds
     cpu time   0.01 seconds

STATUS=SUCCESSFUL PHASE=2 OBJECTIVE=1 P_FEAS=YES D_FEAS=YES PHASE1_I
TER=13 PHASE2_I TER=1 PHASE3_I TER=0

FIGURE 8.23 (Continued)
Log for %ORMCDM.

terminated. So the solution—satisfying the first five priority levels (i.e., P_1 to P_5)—is an optimal solution of the problem, as shown in Figure 8.23. The values of decision variables v_i show that three warehouses were selected, including warehouse 1 ($v_1 = 1$), warehouse 3 ($v_3 = 1$), and warehouse 4 ($v_4 = 1$). The total cost spent in setting up these three warehouses, holding inventory in the warehouses, and delivering products from the warehouses to their assigned customers is $425,000 with no slack. In addition, the total penalty cost incurred is 0. The optimal solution and its total cost are summarized in Table 8.5.

8.2.9 Exercise

Use the codes developed in this chapter and solve the multiple-criteria decision-making problem in Tables 8.6 and 8.7.
    Solution:

- Create the data in a text file (see "data8_2_exercise.txt").
- Run the following code (see program "sasor_8_2_exercise.sas"):

```
* SAS macro for MCDM: solution to exercise 8.2.;
%let _title = 'MCDM, solution to exercise 8.2';
%let _data = 'c:\sasor\MCDM.txt';
%let _dataMCDM = 'c:/sasor/Data8_2_exercise.txt';
%let _nWareh = 5;
%let _nCustomer = 7;
%let _TotalCost = 45800;
%let _outprimal = outprimal;
%let _outdual = outdual;
%let _outtable = outtable;
%let _myLarge = 100000;
%let _nIterations = 100000;
%let _myupperbnd = 12000;
%ormcdm;
```

VAR	_TYPE_	_STATUS_	_LBOUND_	_VALUE_	_UBOUND_	_PRICE_	_R_COST_
X11	INTEGER	_BASIC_	8445	11000	12000	0	0
X12	INTEGER	_BASIC_	0	9000	12000	0	0
X13	INTEGER	_BASIC_	0	10000	12000	0	0
X14	INTEGER	_ALTER_	0	0	12000	0	0
X15	INTEGER	_BASIC_	0	0	12000	0	0
X16	INTEGER	_ALTER_	0	0	12000	0	0
X17	INTEGER	_ALTER_	0	0	12000	0	0
X21	INTEGER	_ALTER_	0	0	12000	0	0
X22	INTEGER	_ALTER_	0	0	12000	0	0
X23	INTEGER	_DEGEN_	0	0	12000	0	0
X24	INTEGER	_ALTER_	0	0	12000	0	0
X25	INTEGER	_ALTER_	0	0	12000	0	0
X26	INTEGER	_ALTER_	0	0	12000	0	0
X27	INTEGER	_ALTER_	0	0	12000	0	0
X31	INTEGER	_BASIC_	0	1000	12000	0	0
X32	INTEGER	_ALTER_	0	0	12000	0	0
X33	INTEGER	_ALTER_	0	0	12000	0	0
X34	INTEGER	_ALTER_	0	0	12000	0	0
X35	INTEGER	_BASIC_	0	6000	12000	0	0
X36	INTEGER	_BASIC_	0	11000	12000	0	0
X37	INTEGER	_ALTER_	0	0	12000	0	0
X41	INTEGER	_ALTER_	0	0	12000	0	0
X42	INTEGER	_ALTER_	0	0	12000	0	0
X43	INTEGER	_ALTER_	0	0	12000	0	0
X44	INTEGER	_BASIC_	0	8000	12000	0	0
X45	INTEGER	_ALTER_	0	0	12000	0	0
X46	INTEGER	_ALTER_	0	0	12000	0	0
X47	INTEGER	_BASIC_	0	7000	12000	0	0
w1	BINARY		0	0	1	0	1.5
u1	BINARY	_BASIC_	0	0	1	0	0
v1	BINARY	_BASIC_	0	1	1	0	0
w2	BINARY		0	0	1	0	1.25
u2	BINARY	_BASIC_	0	1	1	0	0
v2	BINARY	_UPPER_	0	0	0	0	0
w3	BINARY	_DEGEN_	0	0	1	0	0
u3	BINARY		0	0	1	0	1
v3	BINARY	_BASIC_	0	1	1	0	0
w4	BINARY		0	0	1	0	0.75
u4	BINARY	_DEGEN_	0	0	1	0	0
v4	BINARY	UPPER	1	1	1	0	0

FIGURE 8.24
Results of %ORMCDM, primal output of stage 5.

VAR	_TYPE_	_STATUS_	_LBOUND_	_VALUE_	_UBOUND_	_PRICE_	_R_COST_
X11	INTEGER	_BASIC_	8445	11000	12000	0	0
X12	INTEGER	_BASIC_	0	9000	12000	0	0
X13	INTEGER	_BASIC_	0	10000	12000	0	0
X14	INTEGER	_ALTER_	0	0	12000	0	0
X15	INTEGER	_BASIC_	0	0	12000	0	0
X16	INTEGER	_ALTER_	0	0	12000	0	0
X17	INTEGER	_ALTER_	0	0	12000	0	0
X21	INTEGER	_ALTER_	0	0	12000	0	0
X22	INTEGER	_ALTER_	0	0	12000	0	0
X23	INTEGER	_DEGEN_	0	0	12000	0	0
X24	INTEGER	_ALTER_	0	0	12000	0	0
X25	INTEGER	_ALTER_	0	0	12000	0	0
X26	INTEGER	_ALTER_	0	0	12000	0	0
X27	INTEGER	_ALTER_	0	0	12000	0	0
X31	INTEGER	_BASIC_	0	1000	12000	0	0
X32	INTEGER	_ALTER_	0	0	12000	0	0
X33	INTEGER	_ALTER_	0	0	12000	0	0
X34	INTEGER	_ALTER_	0	0	12000	0	0
X35	INTEGER	_BASIC_	0	6000	12000	0	0
X36	INTEGER	_BASIC_	0	11000	12000	0	0
X37	INTEGER	_ALTER_	0	0	12000	0	0
X41	INTEGER	_ALTER_	0	0	12000	0	0
X42	INTEGER	_ALTER_	0	0	12000	0	0
X43	INTEGER	_ALTER_	0	0	12000	0	0
X44	INTEGER	_BASIC_	0	8000	12000	0	0
X45	INTEGER	_ALTER_	0	0	12000	0	0
X46	INTEGER	_ALTER_	0	0	12000	0	0
X47	INTEGER	_BASIC_	0	7000	12000	0	0
w1	BINARY		0	0	1	0	1.5
u1	BINARY	_BASIC_	0	0	1	0	0
v1	BINARY	_BASIC_	0	1	1	0	0
w2	BINARY		0	0	1	0	1.25
u2	BINARY	_BASIC_	0	1	1	0	0
v2	BINARY	_UPPER_	0	0	0	0	0
w3	BINARY	_DEGEN_	0	0	1	0	0
u3	BINARY		0	0	1	0	1
v3	BINARY	_BASIC_	0	1	1	0	0
w4	BINARY		0	0	1	0	0.75
u4	BINARY	_DEGEN_	0	0	1	0	0
v4	BINARY	_UPPER_	1	1	1	0	0

FIGURE 8.25
Results of %ORMCDM, primal output of stage 6.

TABLE 8.5

Optimal Solution

	Allocation of Products, x_{ij}								
			j						
i	1	2	3	4	5	6	7	v_i	$\sum(fc_ihc_i + dc_i)$
1	11,000	9000	10,000	–	–	–	–	1	$220,000
2	–	–	–	–	–	–	–	0	N/A
3	1000	–	–	–	6000	11,000	–	1	$122,000
4	–	–	–	8000	–	–	7000	1	$83,000
								Total	$425,000

TABLE 8.6

Resource Data for the GP Model

	Unit Delivery Cost ($), dc_{ij}						
	Customer, j						
Warehouse, i	1	2	3	4	5	6	7
1	1	2	4	4	5	7	7
2	3	1	5	3	3	7	6
3	5	3	7	3	2	8	6
4	8	4	8	4	1	8	4
5	2	1	5	4	5	8	7
Amount required by customer j, D_j	12,000	9000	10,000	8000	6000	11,000	7000

TABLE 8.7

Resource Data for the GP Model (Continued)

Warehouse, i	Maximum Throughput of Warehouse i Q_i	Minimum Throughput of Warehouse i q_i	Unit Holding Cost ($) hc_i	Fixed Cost ($) fc_i	Penalty Cost ($) pc_i
1	30,000	6000	2	15,000	3000
2	20,000	4000	3	20,000	5000
3	20,000	4000	4	25,000	6000
4	20,000	4000	5	30,000	7000
5	30,000	6000	3	20,000	4000

Targeted total cost, $TC = \$458,000$; Arbitrary large number, $M = 100,000$.

The following solution is given by SAS:

STATUS = INT_INFEASIBLEPHASE = 3OBJECTIVE = 0.51999999999999
P_FEAS = YES   INT_ITER = 6   INT_FEAS = 0   ACTIVE = 0   INT_
BEST = . PHASE1_ITER = 6 PHASE2_ITER = 3 PHASE_ITER = 7

9

Decision Making and Efficiency Measurement

In this chapter, we present a wide range of optimization problems and demonstrate how SAS/OR® can be applied to solve the problems to optimality. The problems include ordered weighted averaging, efficiency measurement, and productivity measurement.

9.1 Ordered Weighted Averaging (OWA) Operators and Preference Ranking

9.1.1 Concept of OWA

The *ordered weighted averaging (OWA) operators* were introduced by Yager (1988). Since then, several applications of OWA operators have been reported in different areas, such as decision making (Yager 1993; Engemann et al. 1996), expert systems (Kacprzyk 1990), neural networks (Eklund and Klawonn 1992), fuzzy systems and control (Yager and Filev 1992), and search engine (Emrouznejad and Amin 2010). Many more applications of OWA have been recently reported in multiple criteria decision making and preference ranking (Yager 1993, Yager et al 2011, Emrouznejad and Amin 2009, 2010, Emrouznejad 2008, 2010, Amin and Emrouznejad 2006, 2009).

An OWA operator of dimension n is a mapping $F : R^n \rightarrow R$ that is associated with weighting vector W in which $W = [w_1, w_2, ..., w_n]^T$ such that

$$\sum_{i=1}^{n} w_i = 1; w_i \in [0,1]$$

and where

$$F(a_1,...,a_n) = \sum_{j=1}^{n} w_j b_j$$

where b_j is the jth largest element of the collection of the aggregated objects $(a_1,..., a_n)$.

The function value $F(a_1,\ldots,a_n)$ determines the aggregated value of arguments $(a_1,\ldots,a_n)$. A fundamental aspect of the OWA operator is the reordering step, in particular an argument that a_i is *not* associated with a particular weight w_i but rather a weight w_i is associated with a particular ordered position i of the arguments. A known property of the OWA operators is that they include the maximum (Max), minimum (Min), and arithmetic mean operators for the appropriate selection of the vector W:

For $\qquad\qquad W = [1, 0,\ldots, 0]^T, \; F\left(a_1,\ldots, a_n\right) = \underset{i}{Max}\left(a_i\right)$

For $\qquad\qquad W = [0,0,\ldots,0]^T, \; F\left(a_1,\ldots, a_n\right) = \underset{i}{Min}\left(a_i\right)$

For $\qquad\qquad W = \left[\dfrac{1}{n},\dfrac{1}{n},\ldots,\dfrac{1}{n}\right]^T, \; F\left(a_1,\ldots, a_n\right) = \dfrac{1}{n}\sum\limits_{i=1}^{n} a_i$

OWA operators are aggregation operators that satisfy the commutative, monotonicity, and idempotency properties and are bounded by the Max and Min operators:

$$\underset{i}{Min}\left(a_i\right) \le F\left(a_i,\ldots, a_n\right) \le \underset{i}{Max}\left(a_i\right)$$

One important issue in the theory of OWA aggregation is the determination of the associated weights. A number of approaches have been suggested in the literature for obtaining the weights. Recently, Amin and Emrouznejad (2006) introduced a linear program to determine the weights (see also Emrouznejad and Amin [2010]).

Each OWA vector is associated with an orness function, usually refered to as the degree of OWA. Assuming that α = orness(w), the linear programming model shown here can be used to determine the OWA operator.

Model 9.1.1 Standard OWA operators model

$$\text{Minimize } z = \delta \qquad\qquad (9.1.1)$$

subject to

$$Orness\left(W\right) = \alpha = \dfrac{1}{n-1}\sum\limits_{i=1}^{n}(n-i)w_i \quad 0 \le \alpha \le 1 \qquad (9.1.2)$$

$$w_j - w_i + \delta \ge 0 \qquad i = 1, \ldots, n-1; j = i+1, \ldots, n \qquad (9.1.3)$$

$$w_i - w_j + \delta \ge 0 \qquad i = 1, \ldots, n-1; j = i+1, \ldots, n \qquad (9.1.4)$$

$$\sum\limits_{i=1}^{n} w_i = 1 \qquad\qquad (9.1.5)$$

$$w_i \ge 0$$

In this model, it is assumed that the degree of OWA is α and the deviation of w_i to w_j is always equal to δ (for all i and j, $i \neq j$).

9.1.2 Example of OWA

Consider a case associated with six criteria and with degree of $\alpha = 0.3$.

Model 9.1.2 Example of formulation of OWA operators

$$\text{Minimize } z = \delta \qquad (9.1.6)$$

subject to

$$w_1 + 0.8w_2 + 0.6w_3 + 0.4w_4 + 0.2w_5 = 0.3 \qquad (9.1.7)$$

$$w_j - w_i + \delta_{ij} \geq 0 \qquad i = 1, \dots, 5; j = 2, \dots, 6; i \neq j \qquad (9.1.8)$$

$$w_i - w_j + \delta_{ij} \geq 0 \qquad i = 1, \dots, 5; j = 2, \dots, 6; i \neq j \qquad (9.1.9)$$

$$w_1 + w_2 + w_3 + w_4 + w_5 + w_6 = 1 \qquad (9.1.10)$$

$$w_i \geq 0$$

An optimal solution is $w_1 = 0.05556$, $w_2 = 0.05556$, $w_3 = 0.05556$, $w_4 = 0.27778$, $w_5 = 0.27778$, $w_6 = 0.27778$. Now consider a set of 10 alternatives that rate 6 criteria as shown in Table 9.1.

Then the rank for alternative 1 is calculated as follows:

$$\text{Rank for alternative } 1 = 3 \times 0.05556 + 3 \times 0.05556 + 3 \times 0.05556$$

$$+ 2 \times 0.27778 + 2 \times 0.27778 + 1 \times 0.27778$$

$$= 1.888889$$

Similarly, the rank for other alternatives are calculated as demonstrated in Table 9.2

TABLE 9.1

Preference Rating of Six Criteria with 10 Alternatives

Alternative	c1	c2	c3	c4	c5	c6
Alt01	3	2	3	2	3	1
Alt02	2	3	3	2	3	2
Alt03	2	2	3	2	2	1
Alt04	3	2	3	3	3	2
Alt05	2	2	3	2	3	1
Alt06	3	2	3	2	3	1
Alt07	1	2	3	2	3	2
Alt08	1	2	3	2	3	1
Alt09	1	2	2	2	3	2
Alt10	1	2	2	3	3	1

TABLE 9.2

Rank of Each Alternative
Using OWA Factor

Alternative	Rank
Alt01	1.888888889
Alt02	2.166666667
Alt03	1.777777778
Alt04	2.444444444
Alt05	1.833333333
Alt06	1.888888889
Alt07	1.833333333
Alt08	1.555555556
Alt09	1.777777778
Alt10	1.555555556

9.1.3 OROWA: SAS Code for OWA

OROWA is a macro that computes OWA using model 9.1 (See program "sasor_9_1.sas"). The OROWA macro first calculates the weights of criteria and then the rank of the alternatives based on the weighted criteria. The procedure used for an OWA problem is PROC OPTMODEL.

Figure 9.1 illustrates the data flow in the OROWA. It shows:

- The matrix of alternatives and criteria that are required for OROWA, in which the importance of each alternative expresses their preference for each criterion
- The macros (%data, %model, and %report)
- The results datasets that are available for print or can be used for further analysis

In the rest of this section, the procedure for implementing preference ranking using OWA in SAS, together with an example, is explained. The OROWA runs three macros: data-handling (%data), model-building (%model), and report-writing (%report).

9.1.4 OROWA: Data-Handling Macro (%data)

This part of OROWA processes the data into a format suitable for use in PROC OPTMODEL. OROWA requires one data set containing the names of the alternatives and criteria and the preference matrix of alternatives for each criterion. The datasets should be .txt files, which are saved as "text tab delimited." The alternatives' names must start with a letter and may contain up to 50 characters. The criteria's names must be listed in the first row of the data file. An example of data file is seen in Figure 9.2.

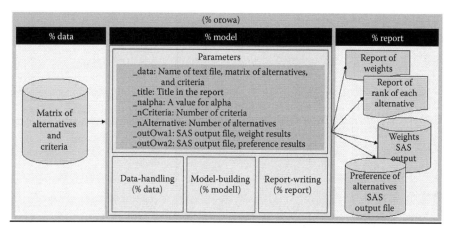

FIGURE 9.1
Data flow in OROWA.

FIGURE 9.2
An example of a dataset, matrix of alternatives, and criteria.

One parameter needs to be set before calling the data macro:

_data: Indicates the name and location of the data file (a text tab delimited file) and contains alternatives' and criteria's matrix

```
* The data-handling macro;
%macro data;
proc import
  datafile = &_data
  out = alts
  dbms = tab
  replace;
  getnames = yes;
run;
%mend data;
```

9.1.5 OROWA: Model-Building Macro (%model)

This part of OROWA processes the data into a format suitable for use in PROC OPTMODEL. Two parameters need to be set before calling the model macro:

_alpha: Identifies the value of alpha

_nCriteria: Identifies the number of criteria

The SAS macro for model-building is as follows:

```
* Macro for calculating OWA operator using LP model in
  Amin-Emrouznejad (AE-OWA);
%macro model;
proc optmodel;
set CRITERIA = 1..&_ncriteria;
set CRITERIA1 = 1..&_ncriteria-1;
set ALTERNATIVE = 1..&_nalternative;
number c{ALTERNATIVE, CRITERIA};
number rankAE{ALTERNATIVE} ;
number rankMIN{ALTERNATIVE} ;
number rankMAX{ALTERNATIVE} ;
number rankMEAN{ALTERNATIVE} ;
number corder{ALTERNATIVE, CRITERIA};
number wMAX{ALTERNATIVE};
number wMIN{ALTERNATIVE};
number wMEAN{ALTERNATIVE};
number ctemp;
number altName{ALTERNATIVE};
number alpha = &_alpha;
number n = &_ncriteria;
var w{CRITERIA} ;
var delta;

read data alts

into [ALTERNATIVE]
{j in CRITERIA} <c[ALTERNATIVE,j] = col("C"||j) > ;

min obj = delta;
con orness: (1/(n-1))* sum{i in CRITERIA}
(n-i)*w[i] = alpha;

con wji{i in 1..n, j in i+1..n }:
w[j]-w[i] + delta >= 0;

con wij{i in 1..n, j in i+1..n }:
w[i]-w[j] + delta >= 0;

con swumw:
sum{i in 1..n} w[i] = 1;
```

```
solve ;
%report;

quit;
%mend model;
```

9.1.6 OROWA: Report-Writing Macro (%report)

The outputs from OROWA include two reports and three SAS output files. Report 1 contains the value of weights and is saved in the SAS dataset "outowa1." Report 2 contains the preference ranking for each alternative and is saved in "outowa2." The user can define appropriate names for each of these datasets before calling %orowa macro as follows:

_outowa1: Identifies the name of the SAS output file for table of weights

_outowa2: Identifies the name of the SAS output file for preference ranking of each alternative

Another parameter needs to be set before calling these macros:

_title: Gives a title in the output of the SAS

```
%macro report;
* MAX-OWA;
for {i in CRITERIA}
if i = 1 then wMAX[i] = 1; else wMAX[i] = 0;

* MIN-OWA;
for {i in CRITERIA}
if i = n then wMIN[i] = 1; else wMIN[i] = 0;

* MEAN-OWA;
for {i in CRITERIA}
wMEAN[i] = 1/n;

*Order the criteria;
for {k in ALTERNATIVE}
 do;
   for {i in CRITERIA}
        for {j in CRITERIA}
        do;
        if c[k,j] < c[k,i] then
            do;
            ctemp = c[k,i]; c[k,i] = c[k,j];c[k,j] = ctemp;
            end;
   end;
 end;
```

```
*Rank by AE-OWA model;
for {k in ALTERNATIVE}
 do;
    rankAE[k] = 0;
    for {i in CRITERIA}
        rankAE[k] = rankAE[k] + c[k,i]*w[i];
    end;
*Rank by MIN-OWA model;
for {k in ALTERNATIVE}
 do;
    rankMIN[k] = 0;
    for {i in CRITERIA}
        rankMIN[k] = rankMIN[k] + c[k,i]*wMIN[i];
    end;
*Rank by MIN-OWA model;
for {k in ALTERNATIVE}
 do;
    rankMAX[k] = 0;
    for {i in CRITERIA}
        rankMAX[k] = rankMAX[k] + c[k,i]*wMAX[i];
    end;
*Rank by MEAN-OWA model;
for {k in ALTERNATIVE}
 do;
    rankMEAN[k] = 0;
    for {i in CRITERIA}
        rankMEAN[k] = rankMEAN[k] + c[k,i]*wMEAN[i];
    end;
* To save the OWA weight in a text file;
file &_owa1;
put 'Criteria,' 'w,' 'wMAX,' 'wMIN,' 'wMEAN';
for {i in CRITERIA}
        do;
        put i ',' w[i] ',' wMAX[i] ',' wMIN[i]',' wMEAN[i];
        end;
* To save the Alternative ranks in a text file;
file &_owa2;
put 'alternative,' 'rankAE,' 'rankMAX,' 'rankMIN,'
'rankMEAN';
for {k in ALTERNATIVE}
    do;
    put k ',' rankAE[k] ',' rankMAX[k] ',' rankMIN[k]','
    rankMEAN[k];
    end;
closefile &_owa1;
closefile &_owa2;
```

```
* To save the OWA weight in a SAS dataset;
proc import
 datafile = &_owa1
 out = &_owaOut1
 dbms = csv
 replace;
 getnames = yes;
run;
proc print; title &_title; run;

* To save the alternative ranks in a SAS dataset;
proc import
 datafile = &_owa2
 out = &_owaOut2
 dbms = csv
 replace;
 getnames = yes;
run;
proc print; title &_title; run;

%mend report;
```

9.1.7 OROWA: Macro (%orowa)

To make the system as user friendly as possible, the %orowa macro combines the data-handling, model-building, and report-writing codes.

```
* The OWA macro for preference setting of a set of
alternatives;
%macro orowa;
  %data;
  %model;
%mend orwa;
```

In this code, the %orowa macro is used to manage all the codes explained earlier, including data-handling, model-building, and report-writing. To get the results, the user needs to set up the parameters and run only one statement:

```
%orowa;
```

9.1.8 Instructions for Using OROWA Macro

This section presents SAS code for the earlier example of preference ranking of 10 alternatives based on 6 criteria as shown in Table 9.1. The data are saved in file "data9_1.txt."

The user needs to set the parameters as required and run the following code:

```
*An example of OWA problem in SAS;
%let _data = 'c:/sasor/Data9_1.txt';
%let _alpha = 0.3;
%let _ncriteria = 6;
%let _nalternative = 10;
%let _owa1 = 'c:/sasor/owaOut1.txt';
%let _owa2 = 'c:/sasor/owaOut2.txt';
%let _owaOut1 = owaOut1;
%let _owaOut2 = owaOut2;
%orowa;
```

This code determines the results based on the specified parameters and the input and output matrices saved in the text files.

9.1.9 Sample Results from OROWA Macro: Output from SAS

The results of running this code are presented in Figures 9.3 and 9.4. This is for the case of $\alpha = 0.3$; the user can choose an appropriate alpha for his or her case.

In Figure 9.4, based on Amin-Emrouznejad ranking (rankAE) method, when $\alpha = 3$, alternatives 4 and 2 are the best because these have the highest score of 2.44444 and 2.1667 respectively, followed by alternatives 1, 6, 5, 7, 3, 9 and finally alternatives 8 and 10.

The reports shown in Figures 9.3 and 9.4 are also saved in SAS datasets &_owaout1 and &_owaout2, respectively.

9.1.10 Exercise

Use the codes developed in this chapter and solve the OWA problem in Table 9.3 with degree of $\alpha = 0.25$.

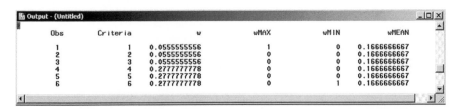

FIGURE 9.3
Results of %orowa, table of weights.

```
Output - (Untitled)                                                        _ □ ×

    Obs     Alternative        rankAE          rankMAX        rankMIN        rankMEAN

     1           1          1.8888888889          3              1         2.3333333333
     2           2          2.1666666667          3              2              2.5
     3           3          1.7777777778          3              1               2
     4           4          2.4444444444          3              2         2.6666666667
     5           5          1.0333333333          3              1         2.1666666667
     6           6          1.8888888889          3              1         2.3333333333
     7           7          1.8333333333          3              1         2.1666666667
     8           8          1.5555555556          3              1               2
     9           9          1.7777777778          3              1               2
    10          10          1.5555555556          3              1               2
```

FIGURE 9.4

Results of %orowa, preference ranking for each alternative.

TABLE 9.3

Preference Rating of Seven Criteria with 15 Alternatives

Alternative	c1	c2	c3	c4	c5	c6	c7
Alt01	5	3	1	3	2	4	1
Alt02	3	2	1	4	2	4	2
Alt03	5	4	5	3	2	4	1
Alt04	2	1	3	3	2	1	5
Alt05	4	3	4	3	1	3	5
Alt06	3	3	2	2	3	2	3
Alt07	2	1	2	4	4	2	2
Alt08	3	3	5	4	4	4	2
Alt09	1	2	5	4	3	3	3
Alt10	1	4	1	5	3	5	1
Alt11	4	2	1	5	2	5	2
Alt12	5	5	4	2	3	3	2
Alt13	1	4	2	4	3	1	2
Alt14	4	1	2	3	3	1	3
Alt15	4	4	2	2	3	3	3

Solution:

- Create the data in a text file (see "data9_1_exercise.txt").
- Run the following code (see program "sasor_9_1_exercise.sas"):

```
* SAS macro for Ordered Weighted Averaging Operators:
solution to exercise 9.1.;
%let _title = 'Ordered Weighted Averaging Operators,
solution to exercise 9.1';
%let _data = 'c:/sasor/Data9_1_exercise.txt';
%let _alpha = 0.25;
```

```
%let _ncriteria = 7;
%let _nalternative = 15;
%let _owa1 = 'c:/sasor/owaOut1.txt';
%let _owa2 = 'c:/sasor/owaOut2.txt';
%let _owaOut1 = owaOut1;
%let _owaOut2 = owaOut2;
%orowa;
```

- The solution in Figure 9.5 is given by SAS.

In Figure 9.6, based on Amin-Emrouznejad ranking method, alternative 8 is the best score for the specified alpha ($\alpha = 0.25$) because it has the highest score (2.8928571429), followed by alternatives 12, 5, 15, and 3. Alternative 10 is the worst, with the lowest score of 1.4642857143.

The reports in Figures 9.5 and 9.6 are saved in SAS dataset &_owaout1 and &_owaout2, respectively.

VIEWTABLE: Work.Owaout1	Criteria	w	wMAX	wMIN	wMEAN
1	1	0.0357142857	1	0	0.1428571429
2	2	0.0357142857	0	0	0.1428571429
3	3	0.0357142857	0	0	0.1428571429
4	4	0.0357142857	0	0	0.1428571429
5	5	0.2857142857	0	0	0.1428571429
6	6	0.2857142857	0	0	0.1428571429
7	7	0.2857142857	0	1	0.1428571429

FIGURE 9.5
Results of %orowa, table of weights.

VIEWTABLE: Work.Owaout2	Alternative	rankAE	rankMAX	rankMIN	rankMEAN
1	1	1.6785714286	5	1	2.7142857143
2	2	1.8928571429	4	1	2.5714285714
3	3	2.3571428571	5	1	3.4285714286
4	4	1.6071428571	5	1	2.4285714286
5	5	2.5714285714	5	1	3.2857142857
6	6	2.1428571429	3	2	2.5714285714
7	7	1.8571428571	4	1	2.4285714286
8	8	2.8928571429	5	2	3.5714285714
9	9	2.25	5	1	3
10	10	1.4642857143	5	1	2.8571428571
11	11	2	5	1	3
12	12	2.6071428571	5	2	3.4285714286
13	13	1.6071428571	4	1	2.4285714286
14	14	1.6071428571	4	1	2.4285714286
15	15	2.5	4	2	3

FIGURE 9.6
Results of %orowa, preference ranking for each alternative.

9.2 Efficiency Measurement Using Data Envelopment Analysis (DEA)

9.2.1 Concept of DEA

Data envelopment analysis (DEA) is a linear programming method for assessing the efficiency and productivity of decision-making units (DMUs). DEA continues to grow in importance as managerial tools become more reliable and can better handle performance measurements of organizations. As a result, new applications with more variables and more complicated models are being introduced. DMUs are units of organizations, such as banks, universities, and hospitals that typically perform the same function. A DMU usually uses a set of inputs (resources) to secure a set of outputs (products) (see Charnes, Cooper, & Rhodes [1978]; Emrouznejad, Parker, & Tavares [2008]; Emrouznejad (2005) and Emrouznejad, De Witte [2010]).

DEA uses linear programming techniques to envelop observed input–output vectors as tightly as possible. One important advantage of DEA is that it allows several inputs and several outputs to be considered at the same time. In this case, efficiency is measured in terms of inputs or outputs along a ray from the origin.

Consider a set of observed DMUs {DMU j; $j = 1, ..., n$} associated with m inputs {x_{ij}; $i = 1, ..., m$} and s outputs, {y_{rj}; $r = 1, ..., s$}. In the method originally proposed by Charnes et al. (1978), the efficiency of the DMU$_{j0}$ is defined as seen in Model 9.2.1.

Model 9.2.1 Standard output-oriented constant returns to scale model

$$\text{Maximize } z = h \tag{9.2.1}$$

subject to

$$\sum_j \lambda_j x_{ij} + S_i^+ = x_{ij0} \quad i = 1, ..., m \tag{9.2.2}$$

$$\sum_j \lambda_j y_{rj} - S_r^- = h y_{rj0} \quad r = 1, ..., s \tag{9.2.3}$$

$$S_i^+, S_r^-, \lambda_j \geq 0$$

where

x_{ij} = The amount of the ith input at DMU j

y_{rj} = The amount of the rth output from DMU j

j_0 = The DMU to be assessed

If h^* is the optimum value of h, then DMU_{j0} is said to be Pareto efficient if and only if $h^* = 1$ and the optimal values of slacks (S_i^+, S_r^-) are 0 for all i and r. This means that no other DMU or combination of DMUs exists that can produce at least the same amount of output as DMU_{j0}, with less for some resources and/or no more for any other resources.

In Model 9.2.1, S_i and S_r represent slack variables. Thus a slack in an input i—that is, $S_i^* > 0$—represents an additional inefficiency use of input i. A slack in an output r—that is, $S_r^* > 0$—represents an additional inefficiency in the production of output r.

DEA Model 9.2.1 is known as *output-oriented model* because it expands output of DMU_{j0} within the production space. It should be solved n times—once for each DMU being evaluated—to generate n optimal values of (h^*, λ^*).

For DMU_{j0}:

- If radial expansion is possible, Model 9.2.1 will yield $h^*_{j0} > 1$
- If radial expansion is not possible, Model 9.2.1 will yield $h^*_{j0} = 1$

The positive elements of the optimal values in λ identify the set of dominating DMUs located on the constructed production frontier, against which the DMU_{j0} is evaluated. This subset of DMUs is called *peers* to DMU_{j0}. From Model 9.2.1, it is clear that the model defines the relative efficiency of a DMU in terms of output maximization. The input-oriented model of DEA can be defined in a similar way.

Model 9.2.2 Standard input-oriented constant returns to scale model

$$\text{Minimize } z = T \tag{9.2.4}$$

subject to

$$\sum_j \lambda_j x_{ij} + S_i^+ = T x_{ij0} \quad i = 1, \ldots, m \tag{9.2.5}$$

$$\sum_j \lambda_j y_{rj} - S_r^- = y_{rj0} \quad r = 1, \ldots, s \tag{9.2.6}$$

$$S_i^+, S_r^-, \lambda_j \geq 0.$$

where

$$x_{ij} = \text{The amount of the } i\text{th input at DMU } j$$
$$y_{rj} = \text{The amount of the } r\text{th output from DMU } j$$
$$j_0 = \text{The DMU to be assessed}$$

Assume that T^* is the optimum value of T. DMU_{j0} is said to be Pareto efficient if and only if $T^* = 1$ and the optimal value of slacks (S_i^+, S_r^-) are 0 for all i and r.

Besides developing DEA in theory, practitioners in a number of fields have quickly recognized that DEA is a useful methodology for measuring productivity and efficiency.

9.2.2 Example of DEA

Table 9.4 shows two inputs and three outputs for efficiency measurement of 10 hospitals.

An input-oriented DEA assessment model for hospital one can be formulated as shown in Model 9.2.3.

Model 9.2.3 Example of formulation of DEA efficiency measurement

$$\text{Minimize } z = H \tag{9.2.7}$$

subject to

$$805\, L_1 + 425\, L_2 + 490\, L_3 + 555\, L_4 + 625\, L_5 + 820\, L_6$$
$$+ 780\, L_7 + 700\, L_8 + 890\, L_9 + 945\, L_{10} \leq 805\, H \tag{9.2.8}$$

$$325\, L_1 + 225\, L_2 + 250\, L_3 + 315\, L_4 + 345\, L_5 + 270\, L_6$$
$$+ 425\, L_7 + 300\, L_8 + 495\, L_9 + 465\, L_{10} \leq 325\, H \tag{9.2.9}$$

TABLE 9.4

A DEA Problem for 10 Hospitals

Hospital	Input1	Input2	Output1	Output2	Output3
1	805	325	330	18	25
2	425	225	140	9	5
3	490	250	160	8	20
4	555	315	220	16	25
5	625	345	310	21	10
6	820	270	260	12	10
7	780	425	430	23	35
8	700	300	590	30	40
9	890	495	810	20	10
10	945	465	510	37	15

$$330\,L_1 + 140\,L_2 + 160\,L_3 + 220\,L_4 + 310\,L_5 + 260\,L_6$$
$$+\,430\,L_7 + 590\,L_8 + 810\,L_9 + 510\,L_{10} \geq 330 \qquad (9.2.10)$$

$$18\,L_1 + 9\,L_2 + 8\,L_3 + 16\,L_4 + 21\,L_5 + 12\,L_6$$
$$+\,23\,L_7 + 30\,L_8 + 20\,L_9 + 37\,L_{10} \geq 18 \qquad (9.2.11)$$

$$25\,L_1 + 5\,L_2 + 20\,L_3 + 25\,L_4 + 10\,L_5 + 10\,L_6$$
$$+\,35\,L_7 + 40\,L_8 + 10\,L_9 + 15\,L_{10} \geq 25 \qquad (9.2.12)$$

$$L_i \geq 0$$

Constraint sets 9.2.8 and 9.2.9 ensure that the inputs of any combination of hospitals are not greater than the inputs of hospital 1, and constraint sets 9.2.10 to 9.2.12 ensure that the outputs of any combination of hospitals are not less than the outputs of hospital 1. Similar formulations should be constructed for other hospitals.

9.2.3 ORDEA: SAS Code for DEA

ORDEA is a macro that implements DEA to measure the efficiency of a set of DMUs (see program "sasor_9_2.sas"). The ORDEA macro can handle both the output maximization model (Model 9.2.1) and the input minimization model (Model 9.2.2). The procedure used for DEA is PROC OPTMODEL.

Figure 9.7 illustrates the data flow in the ORDEA. It shows:

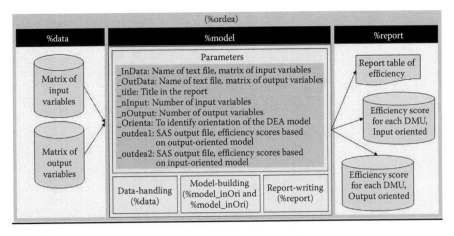

FIGURE 9.7
Data flow in ORDEA.

- The matrix of input variables and the matrix of output variables that are required for ORDEA in which the amount of input levels and output levels for any DMU *j* are specified
- The macros (%data, %model, and %report)
- The results datasets that are available for print or can be used for further analysis

In the rest of this section, the procedure for implementing measurement efficiency using DEA (ORDEA) in SAS, together with an example, is explained. ORDEA runs three macros: data-handling (%data), model-building (%model), and report-writing (%report).

9.2.4 ORDEA: Data-Handling Macro (%data)

This part of ORDEA processes the data into a format suitable for use in PROC OPTMODEL. ORDEA requires two dataset containing the names of DMUs, the matrix of input values, and the matrix of output values. The datasets should be .txt files, which are saved as "text tab delimited." The DMUs' names must start with a letter and may contain up to 50 characters. The variable names must be listed in the first row of the data file. Example data files are seen in Figures 9.8 and 9.9.

Four parameters need to be set before calling the data macro:

_InData: Indicates the name and location of the data file (a text tab delimited file) and contains input matrix

_OutData: Indicates the name and location of the data file (a text tab delimited file) and contains output matrix

_nInput: Identifies the number of input variables

_nOutput: Identifies the number of output variables

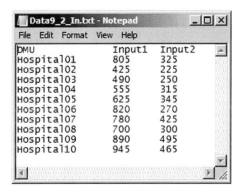

FIGURE 9.8
An example of a dataset, input matrix.

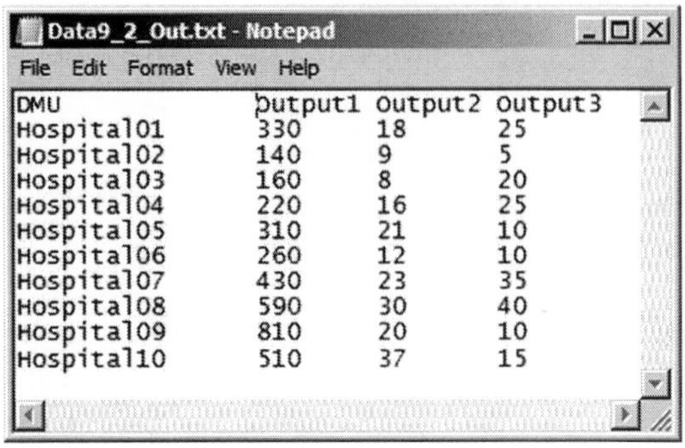

DMU	Output1	Output2	Output3
Hospital01	330	18	25
Hospital02	140	9	5
Hospital03	160	8	20
Hospital04	220	16	25
Hospital05	310	21	10
Hospital06	260	12	10
Hospital07	430	23	35
Hospital08	590	30	40
Hospital09	810	20	10
Hospital10	510	37	15

FIGURE 9.9
An example of a dataset, output matrix.

```
* The data-handling macro;
%macro data;
* Import text tab delimited input variables to SAS data
file;
 proc import
    datafile = &_InData
    out = data9_2_input
    dbms = tab
    replace;
    getnames = yes;
 run;

* Import text tab delimited output variables to SAS data
file;
 proc import
    datafile = &_OutData
    out = data9_2_output
    dbms = tab
    replace;
    getnames = yes;
 run;
 %mend data;
```

9.2.5 ORDEA: Model-Building Macro (%model)

This part of ORDEA formulates an appropriate DEA model in the SAS format that is suitable to PROC OPTMODEL. One parameter needs to be set before calling the model macro:

_Orienta: Identifies the orientation of the DEA mode ("INPUTMIN" for input minimization DEA model and "OUTPUTMAX" for output maximization DEA model).

There are two SAS macros for model-building:

%model_outOri: DEA model with output orientation

%model_inOri: DEA model with input orientation

```
* A DEA macro for CRS output orientation;
%macro model_outOri (j0);
* Starting OPTMODEL Procedure;
proc optmodel;
* Define sets;
set INPUTS = 1..&_nInput;
set OUTPUTS = 1..&_nOutput;
set UNITS = 1..&_nUnits;

* Define variables;
var L{UNITS} > = 0;
var Sin{INPUTS} > = 0 ;
var Sout{OUTPUTS} > = 0 ;
var h;

* Define parameters;
number Xin{UNITS, INPUTS};
number Yout{UNITS,OUTPUTS};
string DMUS{UNITS};
number eff;
string DMUj0;

* Load unit names;
read data data9_2_input
into [_N_]
 DMUS[_N_] = col('dmu');

* Load matrix of input variables;
read data data9_2_input
into [_N_]
{i in INPUTS} < Xin[_N_,i] = col('input'||i) > ;

* Load matrix of output variables;
read data data9_2_output
into [_N_]
{r in OUTPUTS} < Yout[_N_,r] = col('output'||r) > ;

* Define objective function ;
max obj = h;

* Define constraints;
con InpCon_outOri{i in INPUTS}:
sum{j in UNITS} L[j]*Xin[j,i] + Sin[i] = Xin[&j0,i];

con OutCon_outOri{r in OUTPUTS}:
sum{j in UNITS} L[j]*Yout[j,r] - Sout[r] = h*Yout[&j0,r];

* Solve the model;
solve ;
```

```
* Save the efficiency scores in a SAS dataset;
eff = 1/h.sol;
DMUj0 = DMUS[&j0];
create data solj0 from DMUj0 eff;

*End of PROC OPTMODEL;
quit;
%mend model_outOri;
```

```
* A DEA macro for CRS input orientation;
%macro model_inOri (j0);
* Starting OPTMODEL Procedure;
proc optmodel;

* Define sets;
set INPUTS = 1..&_nInput;
set OUTPUTS = 1..&_nOutput;
set UNITS = 1..&_nUnits;

* Define variables;
var L{UNITS} > = 0;
var Sin{INPUTS} > = 0 ;
var Sout{OUTPUTS} > = 0 ;
var h;

* Define parameters;
number Xin{UNITS, INPUTS};
number Yout{UNITS,OUTPUTS};
string DMUS{UNITS};
number eff;
string DMUj0;

* Load unit names;
read data data9_2_input
into [_N_]
 DMUS[_N_] = col('dmu');

* Load matrix of input variables;
read data data9_2_input
into [_N_]
{i in INPUTS} < Xin[_N_,i] = col('input'||i) > ;

* Load matrix of output variables;
read data data9_2_output
into [_N_]
{r in OUTPUTS} < Yout[_N_,r] = col('output'||r) > ;

* Define objective function;
min obj = h;

* Define constraints;
con InpCon_inOri{i in INPUTS}:
```

```
sum{j in UNITS} L[j]*Xin[j,i] + Sin[i] = h*Xin[&j0,i];

con OutCon_inOri{r in OUTPUTS}:
sum{j in UNITS} L[j]*Yout[j,r] - Sout[r] = Yout[&j0,r];

* Solve the model;
solve ;

* Save the efficiency scores in a SAS dataset;
eff = h.sol;
DMUj0 = DMUS[&j0];
create data solj0 from DMUj0 eff ;

*End of PROC OPTMODEL;
quit;
%mend model_inOri;
```

These two macros are written for the case of CRS DEA models, but it is easy to change these macros to calculate the efficiency using VRS DEA models by adding the following constraint to each model:

```
* Define convexity constraints for VRS models;
con Convexity:
sum{j in UNITS} L[j] = 1;
```

9.2.6 ORDEA: Report-Writing Macro (%report)

The outputs from ORDEA include two reports and three SAS output files. Report 1 contains an efficiency table for all DMUs based on the output orientation model, which is saved in the SAS dataset "outdea1." Report 2 contains an efficiency table for all DMUs based on the input orientation model, which is saved in the SAS dataset "outdea2." The user can define appropriate names for each of these datasets before calling %ordea macro:

_outdea1: Identifies the name of the SAS output file for the table of efficiency, output orientation model.

_outdea2: Identifies the name of the SAS output file for the table of efficiency, input orientation model.

Another parameter needs to be set before calling these macros:

_title: Gives a title in the output of the SAS

```
%macro report;
* Delete previously created datasets;
proc datasets nolist;
        delete &_outdea1 &_outdea2;
run;
```

```
* Select the model and execute it for each unit of
assessment;
%do j0 = 1 %to &_nUnits;
%if &_Orienta = "INPUT"
  %then %do;
        %model_inOri (&j0);
        * Save the efficiency scores in a SAS dataset;
        proc datasets nolist;
              append base = &_outdea2 = solj0;
        run;
  %end;
  %else %do;
  %model_outOri (&j0);
        proc datasets nolist;
              append base = &_outdea1 data = solj0;
        run;
  %end;
%end;

%if &_Orienta = "INPUT"
  %then %do;
        * Sort and print the results by efficiency score;
        proc sort data = &_outdea2;
              by eff;
        run;
        proc print data = &_outdea2; title &_title;
        run;

        * Sort and print the results by unit name;
        proc sort data = &_outdea2;
              by DMUj0;
        run;
        proc print data = &_outdea2;
        run;
  %end;
  %else %do;
        * Sort and print the results by efficiency score;
        proc sort data = &_outdea1;
              by eff;
        run;
        proc print data = &_outdea1;
        run;

        * Sort and print the results by unit name;
        proc sort data = &_outdea1;
              by DMUj0;
        run;
        proc print data = &_outdea1; title &_title;
        run;
  %end;

%mend report;
```

9.2.7 ORDEA: Macro (%ordea)

To make the system as user friendly as possible, the %ordea macro processes the data into a format suitable for use in PROC OPTMODEL.

```
*A SAS macro for Data Envelopment Analysis;
%macro ordea;
  %data;
  %report;
%mend ordea;
```

In this code, the %ordea macro is used to manage all the codes explained earlier, including data-handling, model-building, and report-writing. To get the results, user needs to set up the parameters and run only one statement:

```
%ORDEA;
```

9.2.8 Instructions for Using ORDEA Macro

This section presents SAS code for the earlier example of efficiency measurement of 10 hospitals as shown in Table 9.4. The data is saved in files "data9_2_In.txt" and "data9_2_Out.txt."

The user needs to set the parameters as required and run the following code:

```
* Using PROC OPTMODEL for efficiency measurement with
Data Envelopment Analysis (DEA);
option nodate ;
option nonumber ;
%let _title = 'Example 9.2: Efficiency Measurement Using
Data Envelopment Analysis (DEA)';
%let _InData = 'C:\sasor\Data9_2_In.txt';
%let _OutData = 'C:\sasor\Data9_2_Out.txt';
%let _nInput = 2;
%let _nOutput = 3;
%let _nUnits = 10;
%let _Orienta = 'OUTPUT'; *alternative option is 'OUTPUT'
for output orientation;
%let _outdea1 = EffOutOri;
%let _outdea2 = EffInOri;
%ordea;
```

This code determines the results based on the specified parameters and the input and output matrices saved in the text files.

9.2.9 Sample Results from ORDEA Macro: Output from SAS

The results of running this code are presented in Figures 9.10 and 9.11. According to Figure 9.10, hospitals 8 and 9 are the most efficient because their efficiency ratios are equal to 1, followed by hospitals 10, 4, 7, 5, 3, 1, 2, and finally hospital 6.

9.2.10 Exercise

Using the codes developed in this chapter, solve the DEA problem in Table 9.5 and report efficiency of each bank.

Solution:

- Create the data in a text file (see "data9_2_In_exercise.txt" and "data9_2_Out_exercise.txt").
- Run the following code (see program "sasor_9_2_exercise.sas"):

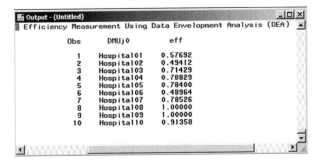

FIGURE 9.10
Results of %ordea, table of efficiency for CRS.

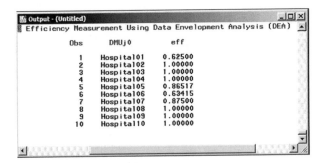

FIGURE 9.11
Sample output report for the case of VRS mode.

TABLE 9.5

A DEA Problem for 20 Bank Branches

Bank	Input 1 Capital	Input 2 Labor	Input 3 Deposits	Output 1 Loans	Output 2 Accounts
Bank01	300	200	1200	12	1300
Bank02	410	440	2000	18	2900
Bank03	930	150	500	12	800
Bank04	200	280	2000	13	1900
Bank05	340	450	800	14	1900
Bank06	300	350	3500	16	2800
Bank07	700	410	2500	21	3000
Bank08	140	120	800	8	700
Bank09	480	270	3500	16	2300
Bank10	400	110	1000	11	900
Bank11	230	180	800	9	900
Bank12	110	500	700	10	1800
Bank13	360	180	2000	12	1400
Bank14	910	200	2500	18	1800
Bank15	760	200	1000	12	1100
Bank16	400	150	1400	10	900
Bank17	750	150	2000	10	900
Bank18	900	350	1500	16	1800
Bank19	840	150	1000	13	1000
Bank20	320	460	800	19	2600

```
* SAS macro for Data Envelopment Analysis: solution to
exercise 9.2.;
%let _title = ' Data Envelopment Analysis, solution to
exercise 9.2';
%let _InData = 'C:\sasor\Data9_2_In_exercise.txt';
%let _OutData = 'C:\sasor\Data9_2_Out_exercise.txt';
%let _nInput = 3;
%let _nOutput = 2;
%let _nUnits = 20;
%let _Orienta = 'OUTPUT'; *alternative option is 'OUTPUT'
for output orientation;
%let _outdea1 = EffOutOri_CRS;
%let _outdea2 = EffInOri_CRS;
%ordea;
```

The following solution is given by SAS:

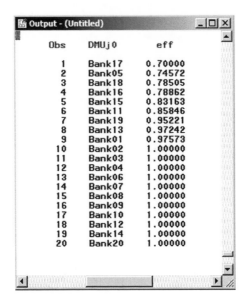

FIGURE 9.12
Results of %ordea, table of efficiency for CRS.

9.3 Productivity Measurement Using Malmquist Index

9.3.1 Concept of Malmquist Index

The *Malmquist index* (Färe et al. 1992) is a productivity measure and can be decomposed to technical change and efficiency change. The concept of the Malmquist productivity index is illustrated by Figure 9.13. In this figure, a production frontier represents the efficiency level of output y and can be produced from a given level of input x. This figure only represents a single-input–single-output case, but it can be extended to multiinput and multioutput in the framework of defining DEA models. The assumption made is that the frontiers can shift over time. The frontiers thus obtained in the present, t, and future, $t + 1$, time periods are labeled accordingly. Therefore, when inefficiency is assumed to exist, the relative movement of any given operational unit over time will depend on both its position relative to the corresponding frontier (efficiency change) and the position of the frontier itself (technical change). If inefficiency is ignored, then the productivity growth over time will be unable to distinguish between improvements that derive from an operational unit catching up to its own frontier and those that result from the frontier itself shifting up over time.

Now assume that $A(t)$ represents an input–output bundle for some given operational unit in period t. Thus, an input-based measure of efficiency can

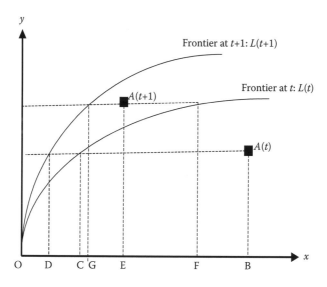

FIGURE 9.13
Malmquist productivity index and its decomposition.

be deduced by the horizontal distance ratio OC/OB. That is, inputs can be reduced to make production technically efficient with respect to the frontier in period t. By comparison and with respect to the same frontier, in period t an input-based measure for operational unit $A(t + 1)$ can be defined with the ratio of OF/OE. Because the frontier has shifted, OF/OE exceeds unity, even though $A(t + 1)$ is technically inefficient when compared with the period $t + 1$ frontier.

With using the Malmquist input-oriented productivity index, it is possible to decompose this total productivity change between the two periods into technical change and efficiency change. Note that some researchers use the input-oriented measures of the Malmquist index (e.g., Funkuyama 1995) but many others use the output orientation of the Malmquist index. The input-based Malmquist productivity index could be formulated as:

$$M_i^{t+1}\left(x^t, y^t, x^{t+1}, y^{t+1}\right) = \left[\frac{D_i^t\left(x^{t+1}, y^{t+1}\right)}{D_i^t\left(x^t, y^t\right)} \times \frac{D_i^{t+1}\left(x^{t+1}, y^{t+1}\right)}{D_i^{t+1}\left(x^t, y^t\right)}\right]^{1/2}$$

where D_i is the input distance function and $M_i^{t+1}(x^t, y^t, x^{t+1}, y^{t+1})$ is the productivity of the most recent production unit—that is, $A(t + 1)$—using period $t + 1$ technology relative to the earlier production unit—that is, $A(t)$—with respect to t technology. A value greater than unity will indicate positive total factor productivity growth between the two periods.

Following Färe et al. (1994), an equivalent way of writing this index is $M = \Delta TECH \times \Delta EFF$ where:

$$EFF = \frac{D_i^{t+1}\left(x^{t+1}, y^{t+1}\right)}{D_i^t\left(x^t, y^t\right)}$$

$$TECH = \left[\frac{D_i^t\left(x^{t+1}, y^{t+1}\right)}{D_i^{t+1}\left(x^{t+1}, y^{t+1}\right)} \times \frac{D_i^t\left(x^t, y^t\right)}{D_i^{t+1}\left(x^t, y^t\right)}\right]^{1/2}$$

In this view, M, the Malmquist total factor productivity index, is the product of a measure of technical progress, $\Delta TECH$, as measured by shifts in a frontier at period $t + 1$ and period t (average geometrically) and a change in efficiency, ΔEFF, over the same period.

To calculate these indexes, it is necessary to solve several sets of linear programming problems as presented in Models 9.3.1 to 9.3.4. Assume that there are n DMUs and that each DMU consumes varying amounts of m different inputs to produce s outputs in each period t. Therefore the jth DMU in period t is represented by the vectors (x_j^t, y_j^t). The purpose is to construct a nonparametric envelopment frontier over the data points such that all observed DMUs lie on or below the production frontier. The calculation exploits the fact that the input distance functions (D_i) used to construct the Malmquist index are the reciprocals of the Farrell (1957) input-oriented technical efficiency measure. The first two linear programs (Models 9.3.1 and 9.3.2) are used to evaluate the technology and observation from the same period, and the solution value is less than or equal to unity. The last two linear programs (Models 9.3.3 and 9.3.4) are employed to construct the reference technology from data in one period, whereas the observation to be evaluated is from another period. Assuming that constant returns to scale, the following four linear programs are used to calculate the Malmquist index and its components.

Model 9.3.1 Linear programming model for calculation of Malmquist productivity index and its components, T_{11}

$$\left[D_i^t\left(x_t,\ y_t\right)\right]^{-1} = Min\ \phi \tag{9.3.1}$$

subject to

$$\sum_j \lambda_j x_{ij}^t \leq x_{ij0}^t \quad i = 1, \ldots, m \tag{9.3.2}$$

$$\sum_j \lambda_j y_{rj}^t \geq y_{rj0}^t \quad r = 1, \ldots, s \tag{9.3.3}$$

$$\lambda_j \geq 0$$

Model 9.3.2 Linear programming model for calculation of Malmquist productivity index and its components, T_{22}

$$\left[D_i^{t+1}(x_{t+1}, y_{t+1})\right]^{-1} = Min \; \phi \tag{9.3.4}$$

subject to

$$\sum_j \lambda_j x_{ij}^{t+1} \le x_{ij0}^{t+1} \quad i = 1, \ldots, m \tag{9.3.5}$$

$$\sum_j \lambda_j y_{rj}^{t+1} \ge y_{rj0}^{t+1} \quad r = 1, \ldots, s \tag{9.3.6}$$

$$\lambda_j \ge 0$$

Model 9.3.3 Linear programming model for calculation of Malmquist productivity index and its components, T_{12}

$$\left[D_i^{t}(x_{t+1}, y_{t+1})\right]^{-1} = Min \; \phi \tag{9.3.7}$$

subject to

$$\sum_j \lambda_j x_{ij}^{t} \le x_{ij0}^{t+1} \quad i = 1, \ldots, m \tag{9.3.8}$$

$$\sum_j \lambda_j y_{rj}^{t} \ge y_{rj0}^{t+1} \quad r = 1, \ldots, s \tag{9.3.9}$$

$$\lambda_j \ge 0$$

Model 9.3.4 Linear programming model for calculation of Malmquist productivity index and its components, T_{21}

$$\left[D_i^{t+1}(x_t, y_t)\right]^{-1} = Min \; \phi \tag{9.3.10}$$

subject to

$$\sum_j \lambda_j x_{ij}^{t+1} \le x_{ij0}^{t} \quad i = 1, \ldots, m \tag{9.3.11}$$

$$\sum_j \lambda_j y_{rj}^{t+1} \ge y_{rj0}^{t} \quad r = 1, \ldots, s \tag{9.3.12}$$

$$\lambda_j \ge 0$$

By solving these linear programming models, it is possible to provide four efficiency and productivity indexes for each observed DMU. Regarding change in efficiency, technical efficiency increases if and only if the optimum ΔEFF is greater than 1 and ΔEFF can be obtained by solving the first two linear programming models. An interpretation of the technological change is that technical progress has occurred if $\Delta TECH$ is greater than 1. In contrast, efficiency decreases if and only if the optimum ΔEFF is less than 1 and EFF can be obtained by solving the first two linear programming models; this can be interpreted as technical regress has occurred if $\Delta TECH$ is less than 1.

9.3.2 Example of Malmquist Index

Table 9.6 shows two inputs and two outputs for productivity measurement of six DMUs in 2 years.

An input-oriented Malmquist index for Unit1 needs to solve four linear programming as formulated in Models 9.3.5 to 9.3.8.

Model 9.3.5 Example of DEA efficiency measurement formulation, both technology and assessment unit in period 2007

$$\text{Minimize } z = T_{11} \tag{9.3.13}$$

subject to

$$15\,L_1 + 40\,L_2 + 32\,L_3 + 52\,L_4 + 35\,L_5 + 32\,L_6 \leq 15\,T_{11} \tag{9.3.14}$$

$$2\,L_1 + 7\,L_2 + 12\,L_3 + 20\,L_4 + 12\,L_5 + 7\,L_6 \leq 2\,T_{11} \tag{9.3.15}$$

$$14\,L_1 + 14\,L_2 + 42\,L_3 + 28\,L_4 + 19\,L_5 + 14\,L_6 \geq 14 \tag{9.3.16}$$

$$3.5\,L_1 + 21\,L_2 + 10.5\,L_3 + 42\,L_4 + 25\,L_5 + 15\,L_6 \geq 3.5 \tag{9.3.17}$$

$$L_i \geq 0$$

TABLE 9.6

A Malmquist Problem for Six DMUs

DMU	2007				2008			
	IN1	IN2	OUT1	OUT2	IN1	IN2	OUT1	OUT2
Unit1	15	2	14	3.5	10	1.5	17	2.5
Unit2	40	7	14	21	45	5.6	16	22
Unit3	32	12	42	10.5	35	11	40	10
Unit4	52	20	28	42	50	27	28	30
Unit5	35	12	19	25	30	14	19	25
Unit6	32	7	14	15	38	9	13	12

Model 9.3.6 Example of DEA efficiency measurement formulation, both technology and assessment unit in period 2008

$$\text{Minimize } z = T_{22} \tag{9.3.18}$$

subject to

$$10\,L_1 + 45\,L_2 + 35\,L_3 + 50\,L_4 + 30\,L_5 + 38\,L_6 \leq 10\,T_{22} \tag{9.3.19}$$

$$1.5\,L_1 + 5.6\,L_2 + 11\,L_3 + 27\,L_4 + 14\,L_5 + 9\,L_6 \leq 1.5\,T_{22} \tag{9.3.20}$$

$$17\,L_1 + 16\,L_2 + 40\,L_3 + 28\,L_4 + 19\,L_5 + 13\,L_6 \geq 17 \tag{9.3.21}$$

$$2.5\,L_1 + 22\,L_2 + 10\,L_3 + 30\,L_4 + 25\,L_5 + 12\,L_6 \geq 2.5 \tag{9.3.22}$$

$$L_i \geq 0$$

Model 9.3.7 Example of DEA efficiency measurement formulation, technology in period 2007 and assessment unit in period 2008

$$\text{Minimize } z = T_{12} \tag{9.3.23}$$

subject to

$$15\,L_1 + 40\,L_2 + 32\,L_3 + 52\,L_4 + 35\,L_5 + 32\,L_6 \leq 10\,T_{12} \tag{9.3.24}$$

$$2\,L_1 + 7\,L_2 + 12\,L_3 + 20\,L_4 + 12\,L_5 + 7\,L_6 \leq 1.5\,T_{12} \tag{9.3.25}$$

$$14\,L_1 + 14\,L_2 + 42\,L_3 + 28\,L_4 + 19\,L_5 + 14\,L_6 \geq 17 \tag{9.3.26}$$

$$3.5\,L_1 + 21\,L_2 + 10.5\,L_3 + 42\,L_4 + 25\,L_5 + 15\,L_6 \geq 2.5 \tag{9.3.27}$$

$$L_i \geq 0$$

Model 9.3.8 Example of DEA efficiency measurement formulation technology in period 2008 and assessment unit in period 2007

$$\text{Minimize } z = T_{21} \tag{9.3.28}$$

subject to

$$10\,L_1 + 45\,L_2 + 35\,L_3 + 50\,L_4 + 30\,L_5 + 38\,L_6 \leq 15\,T_{21} \tag{9.3.29}$$

$$1.5\,L_1 + 5.6\,L_2 + 11\,L_3 + 27\,L_4 + 14\,L_5 + 9\,L_6 \leq 2\,T_{21} \tag{9.3.30}$$

$$17\,L_1 + 16\,L_2 + 40\,L_3 + 28\,L_4 + 19\,L_5 + 13\,L_6 \geq 14 \tag{9.3.31}$$

$$2.5\,L_1 + 22\,L_2 + 10\,L_3 + 30\,L_4 + 25\,L_5 + 12\,L_6 \geq 3.5 \qquad (9.3.32)$$

$$L_i \geq 0$$

By solving these four linear programs, the Malmquist index can be obtained.

9.3.3 ORMALM: SAS Code for Malmquist Index

The ORMALM macro introduced in this section provides a powerful management tool for assessing both efficiency and productivity of organizations in an SAS system (see program "sasor_9_3.sas"). The program can handle both input minimization and output maximization and can also calculate the input and output Malmquist index and its components. There are several parameters to enhance the model. The user can select the desired parameters according to the particular model that is required. Users familiar with SAS can add their own features to build other DEA models. Users not familiar with SAS need only to run the program with their model specification before running the system.

The ORMALM macro requires several initial datasets that contain the names of the variables, value for observed units in each period. The data describing inputs and outputs must be formatted so that variables appear in columns and units in rows. The data must be saved as .txt (tab delimited) files. The program has the ability to accommodate an unlimited number of inputs and outputs with an unlimited number of DMUs. The only limitation is the memory of computer used to run the macro.

The ORMALM macro then converts datasets to a selected DEA model. Based on the data and parameters specified in ORMALM, the code first creates the usual linear program. Then it uses PROC OPTMODEL to solve the model. The results will be transferred to report files.

ORMALM produces a table of efficiencies of DMUs. It also supplies other valuable information, including lambda and slack values in the primal DEA model and weights in the dual DEA model. This information is very useful for analyzing the inefficient units, where the source of inefficiency comes from, and how to improve an inefficient unit to the desired level.

Figure 9.14 illustrates the data flow in the ORMALM. It shows:

- The matrix of input–output variables and the names of input–output variables that are required for ORMALM, in which the amount of input levels and output levels in each year and for any DMU j are specified
- The macros (%data, %model, and %report)
- The results datasets that are available for print or can be used for further analysis

In the rest of this section, the procedure for implementing efficiency measurment using the Malmquist index (ORMALM) in SAS, together with

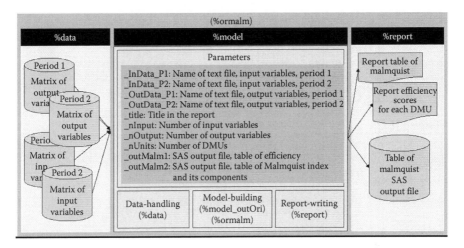

FIGURE 9.14
Data flow in ORMALM.

an example, is explained. ORMALM runs six macros: two for data-handling (%data1 and %data2), three for model-building (%model, %duality, and %dea), and one for report-writing (%report).

9.3.4 ORMALM: Data-Handling Macro (%data)

This part of ORMALM processes the data into a format suitable for use in PROC OPTMODEL. The ORMALM macro requires four datasets containing the amount of input–output variables for each period. The datasets should be .txt files, which are saved as "text tab delimited." The DMUs' names must start with a letter and may contain up to 50 characters. The variable names must be listed in the first row of the data file. Example data files are seen in Figures 9.15 to 9.18.

Six parameters need to be set before calling the data macro:

_InData_P1: Indicates the name and location of the data file (a text tab delimited file) and contains input variables for period 1

_InData_P2: Indicates the name and location of the data file (a text tab delimited file) and contains input variables for period 2

_OutData_P1: Indicates the name and location of the data file (a text tab delimited file) and contains output variables for period 1

_OutData_P2: Indicates the name and location of the data file (a text tab delimited file) and contains output variables for period 2

_nInput: Indicates the number of input variables

_nOutput: Indicates the number of output variables

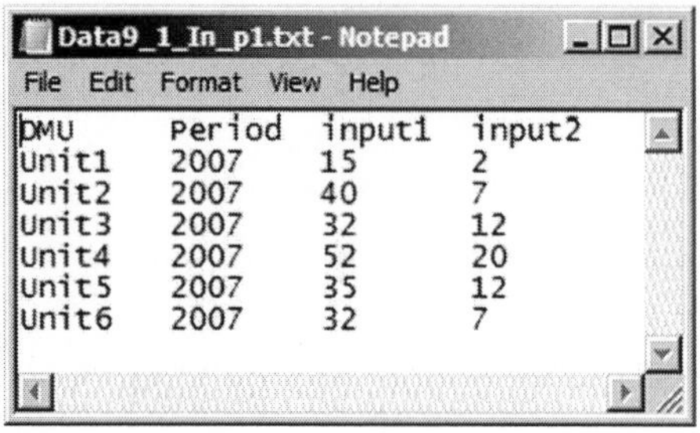

Data9_1_In_p1.txt - Notepad

DMU	Period	input1	input2
Unit1	2007	15	2
Unit2	2007	40	7
Unit3	2007	32	12
Unit4	2007	52	20
Unit5	2007	35	12
Unit6	2007	32	7

FIGURE 9.15
An example of a dataset, input variables, and period 1.

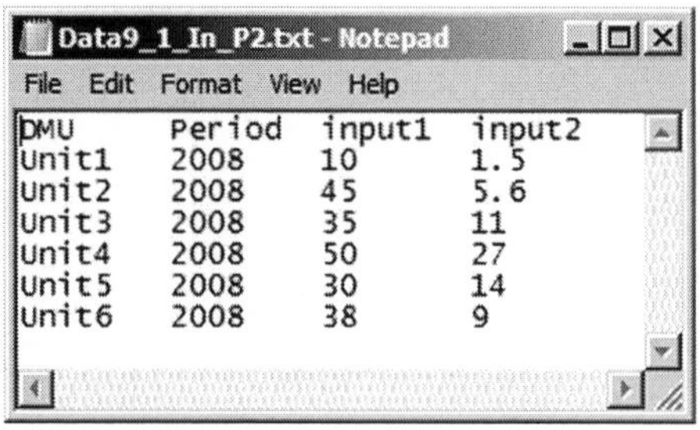

Data9_1_In_P2.txt - Notepad

DMU	Period	input1	input2
Unit1	2008	10	1.5
Unit2	2008	45	5.6
Unit3	2008	35	11
Unit4	2008	50	27
Unit5	2008	30	14
Unit6	2008	38	9

FIGURE 9.16
An example of a dataset, input variables, and period 2.

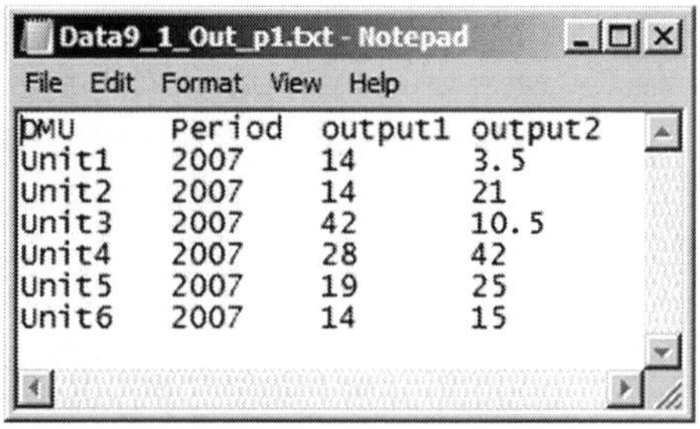

Data9_1_Out_p1.txt - Notepad

DMU	Period	output1	output2
Unit1	2007	14	3.5
Unit2	2007	14	21
Unit3	2007	42	10.5
Unit4	2007	28	42
Unit5	2007	19	25
Unit6	2007	14	15

FIGURE 9.17
An example of a dataset, output variables, and period 1.

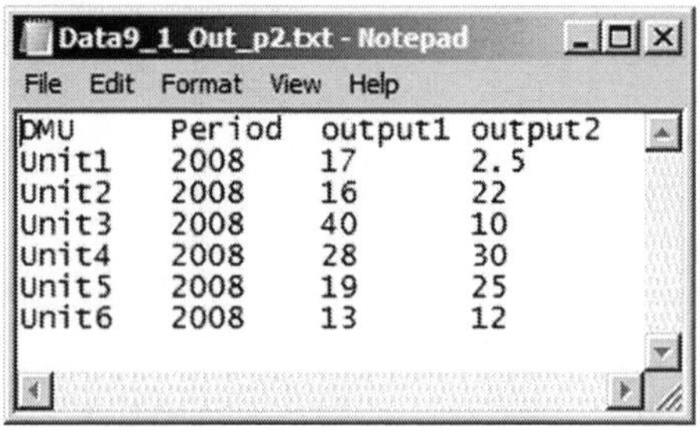

Data9_1_Out_p2.txt - Notepad

DMU	Period	output1	output2
Unit1	2008	17	2.5
Unit2	2008	16	22
Unit3	2008	40	10
Unit4	2008	28	30
Unit5	2008	19	25
Unit6	2008	13	12

FIGURE 9.18
An example of a dataset, output variables, and period 2.

```
%macro importdata (txtFile, sasFile);
  proc import
         datafile = &txtFile
         out = &sasFile
         dbms = tab
         replace;
         getnames = yes;
  run;
%mend importdata;

* The data-handling macro;
%macro data;
* Import text tab delimited input variables to SAS data
file, period1;
%importdata(&_InData_P1,date9_3_input_P1);
* Import text tab delimited output variables to SAS data
file, period1;
%importdata(&_OutData_P1,date9_3_output_P1);
* Import text tab delimited input variables to SAS data
file, period2;
%importdata(&_InData_P2,date9_3_input_P2);
* Import text tab delimited output variables to SAS data
file, period2;
%importdata(&_OutData_P2,date9_3_output_P2);
%mend data;
```

9.3.5 ORMALM: Model-Building Macro (%model_outOri)

This part of ORMALM formulates an appropriate DEA model in the SAS format that is suitable for use in PROC OPTMODEL.

SAS macro for model-building is as follows:

```
* A DEA macro for CRS output orientation;
%macro model_outOri ( per1, per2,j0);
* Starting OPTMODEL Procedure;
proc optmodel;
* Define sets;
set INPUTS = 1..&_nInput;
set OUTPUTS = 1..&_nOutput;
set UNITS = 1..&_nUnits;
set PERIODS = 1..2;
```

```
* Define variables;
var L{UNITS} >= 0;
var Sin{INPUTS} >= 0 ;
var Sout{OUTPUTS} >= 0 ;
var h;

* Define parameters;
number Xin{PERIODS,UNITS,  INPUTS};
number Yout{PERIODS,UNITS,OUTPUTS};
string DMUS{UNITS};
number eff;
string DMUj0;
number p1;
number p2;

* Load unit names;
read data date9_3_input_P1
into [_N_]
 DMUS[_N_] = col('dmu');

* Load matrix of input variables;
read data date9_3_input_P1
into [_N_]
{i in INPUTS} < Xin[1,_N_,i] = col('input'||i) > ;

* Load matrix of input variables;
read data date9_3_input_P2
into [_N_]
{i in INPUTS} < Xin[2,_N_,i] = col('input'||i) > ;

* Load matrix of input variables;
read data date9_3_output_P1
into [_N_]
{r in OUTPUTS} < Yout[1,_N_,r] = col('output'||r) > ;

* Load matrix of input variables;
read data date9_3_output_P2
into [_N_]
{r in OUTPUTS} < Yout[2,_N_,r] = col('output'||r) > ;

* Define objective function ;
max obj = h;

* Define constraints;
con InpCon_outOri{i in INPUTS}:
sum{j in UNITS}
L[j]*Xin[&Per1,j,i] + Sin[i] = Xin[&Per2,&j0,i];

con OutCon_outOri{r in OUTPUTS}:
sum{j in UNITS} L[j]*Yout[&Per1,j,r]
- Sout[r] = h*Yout[&Per2,&j0,r];
* Solve the model;
solve ;
```

```
* Save the efficiency scores in a SAS dataset;
eff = 1/h.sol;
DMUj0 = DMUS[&j0];
p1 = &Per1;
p2 = &Per2;
create data solj0 from DMUj0 p1 p2 eff;

*End of PROC OPTMODEL;
quit;
%mend model_outOri;
```

This macro is written for finding efficiency scores with output orientation. It is easy to change the macro to an input-orientation model by replacing the objective function and constraints to:

```
* Define objective function;
min obj = h;

* Define constraints;
con InpCon_InOri{i in INPUTS}:
sum{j in UNITS}
L[j]*Xin[&Per1,j,i] + Sin[i] = h*Xin[&Per2,&j0,i];

con OutCon_InOri{r in OUTPUTS}:
sum{j in UNITS} L[j]*Yout[&Per1,j,r]
- Sout[r] = Yout[&Per2,&j0,r];
```

9.3.6 ORMALM: Report-Writing Macro (%report)

The outputs from ORMALM include one report and one SAS output files. The report contains a table of efficiency, the Malmquist index, and decomposition of Malmquist for all DMUs that are saved in the SAS dataset "outMalm1."

_outMalm1: Identifies the name of the SAS output file for table of efficiency and Malmquist index

Another parameter needs to be set before calling this macro:

_title: Gives a title in the output of the SAS

```
* Macro for reporting the results;
%macro report(per1,per2);
* Select the model and execute it for each unit of
assessment;
%do j0 = 1 %to &_nUnits;
        %model_outOri (&per1,&per2,&j0);
            * Save the results for report-writing;
```

```
        proc datasets nolist;
                append base = &_outMalm1 data = solj0;
        run;
  %end;
%mend report;
```

9.3.7 ORMALM: Macro (%ormalm)

To make the system as user friendly as possible, the %ormalm macro combines the data-handling, model-building, and report-writing codes.

```
%macro ormalm;
 %data;
 * Delete previously created datasets;
 proc datasets nolist;
        delete &_outMalm1;
 run;
%report(1,1);
%report(2,2);
%report(1,2);
%report(2,1);
data &_outMalm2;
merge &_outMalm1(WHERE = (p1 = 1 and p2 = 1)
        RENAME = (eff = DEA11))
        &_outMalm1(WHERE = (p1 = 1 and p2 = 2)
        RENAME = (eff = DEA12))
        &_outMalm1(WHERE = (p1 = 2 and p2 = 1)
        RENAME = (eff = DEA21))
        &_outMalm1(WHERE = (p1 = 2 and p2 = 2)
        RENAME = (eff = DEA22));
                by dmuj0;
                drop p1 p2;
 run;
data &_outMalm2(drop = DEA21 DEA12
rename = (DEA11 = EffPeriod1 DEA22 = EffPeriod2));
        set &_outMalm2;
        EffChange = DEA22/DEA11;
        TechChange = sqrt((DEA12*DEA11)/(DEA22*DEA21));
        MalmIndex = EffChange*TechChange;
 run;
proc print; run;
%mend ormalm;
```

In this code, the %ormalm macro is used to manage all the codes explained earlier, including data-handling, model-building, and report-writing. To get the results, the user needs to set up the parameters and run only one statement:

```
%ORMALM;
```

9.3.8 Instructions for Using ORMALM Macro

This section presents SAS code for the earlier example of efficiency measurement of six DMUs in two periods of 2007 and 2008 as shown in Table 9.6. The data are saved in files "data9_3_In_P1.txt", "data9_3_In_P2.txt", "data9_3_Out_P1.txt", and "data9_3_Out_P2.txt."

The user needs to set the parameters as required and run the following code:

```
* Parameter definitions and running SAS/MALM;
%let _title = 'Example 9.3. Malmquist index and its
components';
%let _libname = 'c:/sasor/ormalm'; * Name of directory;
%let _Data1 = 'C:/sasor/ormalm/Data9_3.txt';
%let _Data2 = 'C:/sasor/ormalm/Data9_3_var.txt';
%let _Period1 = 2007; * First period;
%let _Period2 = 2008; * second period;
%let _outMalm1 = outMalm1;
%let _outMalm2 = outMalm2;
%let _Orienta = 'INPUTMIN'; * Alternative selection is
'OUTPUTMAX';
%ormalm;
```

This code determines the results based on the specified parameters and the input and output matrix saved in the text files.

9.3.9 Sample Results from ORMALM Macro: Output from SAS

The results of running this code are presented in Figure 9.19; columns Eff Period1 and Eff Period2 refer to the efficiency score for periods 1 and 2, respectively. The efficiency change, technical change, and Malmquist productivity index are reported in the last three columns. The Malmquist index of Unit1 is the highest (1.51568), which means that its improvement rate in productivity is the highest, followed by Unit2, Unit5, Unit3, Unit4, and finally Unit6.

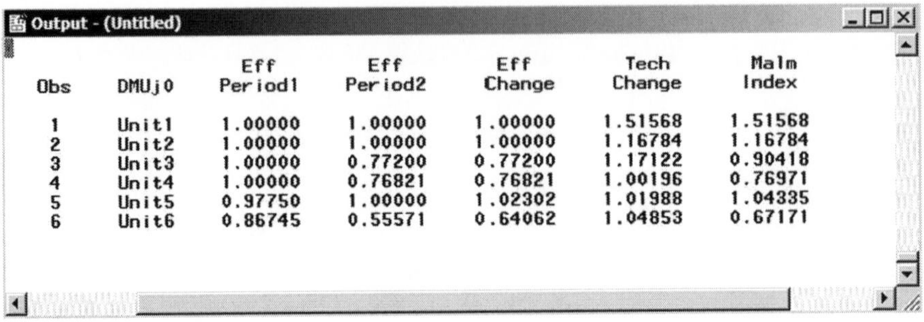

Output - (Untitled)

Obs	DMUj0	Eff Period1	Eff Period2	Eff Change	Tech Change	Malm Index
1	Unit1	1.00000	1.00000	1.00000	1.51568	1.51568
2	Unit2	1.00000	1.00000	1.00000	1.16784	1.16784
3	Unit3	1.00000	0.77200	0.77200	1.17122	0.90418
4	Unit4	1.00000	0.76821	0.76821	1.00196	0.76971
5	Unit5	0.97750	1.00000	1.02302	1.01988	1.04335
6	Unit6	0.86745	0.55571	0.64062	1.04853	0.67171

FIGURE 9.19
Results of %ormalm, table of efficiency, and Malmquist index.

TABLE 9.7
A Malmquist Problem for 10 DMUs

	2010				2011			
DMU	IN1	IN2	OUT1	OUT2	IN1	IN2	OUT1	OUT2
DMU1	25	40	30	20	21	50	35	10
DMU2	65	100	25	60	60	90	30	50
DMU3	70	150	60	30	60	120	70	20
DMU4	80	190	70	40	90	150	50	50
DMU5	65	150	40	50	70	120	45	45
DMU6	70	100	40	40	80	120	30	30
DMU7	90	120	50	55	75	100	40	45
DMU8	80	110	55	60	65	110	50	60
DMU9	65	90	45	55	80	150	45	45
DMU10	70	100	40	65	90	145	50	65

9.3.10 Exercise

Use the codes developed in this chapter and calculate the Malmquist index and its components for the 10 DMUs with 2 input and 2 output variables found in Table 9.7.

Solution:

- Create the data in a text file (see "data9_3_In_p1_exercise.txt", "data9_3_In_p2_exercise.txt", " data9_3_Out_p1_exercise.txt", and "data9_3_Out_p2_exercise.txt").
- Run the following code (see program "sasor_9_3_exercise.sas").

```
Output - (Untitled)                                                    _ □ ×
                    Eff        Eff        Eff       Tech       Malm
     Obs   DMUj0   Period1    Period2    Change     Change     Index

      1    DMU01   1.00000    1.00000    1.00000    1.02869    1.02869
      2    DMU02   0.99408    1.00000    1.00595    0.90705    0.91245
      3    DMU03   0.71429    0.83333    1.16667    1.12802    1.31603
      4    DMU04   0.72917    0.67708    0.92857    0.98092    0.91086
      5    DMU05   0.85619    0.76172    0.88966    0.99953    0.88924
      6    DMU06   0.69333    0.50781    0.73242    0.85456    0.62590
      7    DMU07   0.77222    0.85469    1.10679    0.87871    0.97255
      8    DMU08   0.92121    1.00000    1.08553    0.95040    1.03169
      9    DMU09   1.00000    0.64512    0.64512    0.94749    0.61124
     10    DMU10   1.00000    0.81773    0.81773    0.91359    0.74707
```

FIGURE 9.20
Results of %ormalm, table of Malmquist index and its components, for CRS.

```
* SAS macro for Malmquist Index and its components:
solution to exercise 9.3.;
%let _title = ' Malmquist Index and its components,
solution to exercise 9.3';
%let _InData_P1 = 'C:\sasor\Data9_3_In_P1_exercise.txt';
%let _OutData_P1 = 'C:\sasor\Data9_3_Out_P1_exercise.txt';
%let _InData_P2 = 'C:\sasor\Data9_3_In_P2_exercise.txt';
%let _OutData_P2 = 'C:\sasor\Data9_3_Out_P2_exercise.txt';
%let _nInput = 2;
%let _nOutput = 2;
%let _nUnits = 10;
%let _outMalm1 = outMalm1;
%let _outMalm2 = outMalm2;
%ormalm;
```

The solution in Figure 9.20 is given by SAS.

Appendix 1: Syntax of PROC IMPORT and PROC EXPORT

This appendix gives a brief introduction to import and export datasets to and from SAS to other packages, including text delimited and Excel files (see also "sasor_import_export.sas").

Import Data from a Comma-Delimited Text File

It is quite easy to read a file that uses a comma as a delimiter using PROC IMPORT in SAS. Assume that the following comma-delimited dataset is stored in a file called "c:/sasor/mydata.csv."

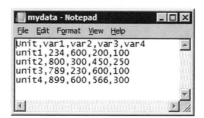

Then the following PROC IMPORT statement will read it in and create a temporary dataset called "mydata."

```
proc import
        datafile = "c:/sasor/mydata.csv"
        out = mydata
        dbms = csv
        replace;
        getnames = yes;
run;
```

As you can see in the output below, the data was read properly. The names of variables are read from the first row. If the names of variables are not given in the first row, then SAS by default creates VAR1-VARn when variables names are not present in the raw data file.

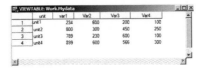

Import Data from a Tab-Delimited Text File

Similar to the earlier example, it is easy to read a data file that employs a tab as a delimiter using PROC IMPORT in SAS. Assume that the following tab-delimited dataset is stored in a file called "c:/sasor/mydata.txt."

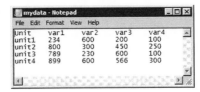

Then the following PROC IMPORT statement will read it in and create a temporary dataset called "mydata."

```
proc import
       datafile = "c:/sasor/mydata.txt"
       out = mydata
       dbms = tab
       replace;
       getnames = yes;
run;
```

Import Data from a Space-Delimited Text File

Assume that the following space-delimited dataset is stored in a file called "c:/sasor/mydataspace.txt."

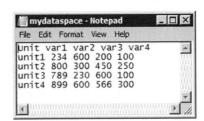

Then the following PROC IMPORT statement will read it in and create a temporary data set called "mydata."

```
proc import
        datafile = "c:/sasor/mydataspace.txt"
        out = mydata
        dbms = dlm
        replace;
        getnames = yes;
run;
```

Import Data from an EXCEL File

Assume that the following Excel dataset is stored in a file called "c:/sasor/ mydata.xls":

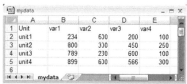

Then the following PROC IMPORT statement will read it in and create a temporary dataset called "mydata":

```
proc import
        datafile = "c:/sasor/mydataspace.txt"
        out = mydata
        dbms = EXCEL2000
        replace;
        getnames = yes;
run;
```

Export Data from SAS to Other Formats

Similarly, data files can be exported to other formats using PROC EXPORT. The following is an example of this procedure:

```
proc export
   data = mydata
   outfile = 'c:/sasor/mydatout.txt'
   dbms = dlm
   replace;
   delimiter = ',';
run;
```

Appendix 2: Syntax of PROC OPTMODEL*

```
PROC OPTMODEL

/*option Statements*/
options ;

*Declaration Statements;
constraint ...;   *(or con);
max ...;
min ...;
number ...;              *(or num);
string ...;              *(or str);
set    ...;
var    ...;

*General statements;
Assignment...;
Call...;
Reset ...;

*Input/Output statements;
Closefile...;
Create data...;
File...;
Print...;
Put...;
Read data...;
Save MPS...;
Save QPS...;

*Loop statements;
Continue ...;
for ...;
do iterative...;
do until ...;
do while ...;
leave ...;

*Control statements;
do ...;
if-then-else...;
stop ...;
```

---

* For full details, see the SAS/OR® User's Guide.

```
*Model statements;
drop ...;
expand ...;
fix ...;
restore ...;
solve ...;
unfix ...;
QUIT;
```

Appendix 3: Further Resources for SAS/OR®

Accessing the SAS/OR®-Related Materials Library

The webpage www.sas-or.com has been implemented with the aim of growing and updating new materials about SAS/OR® tools. It is recommended that users access this website and share their expertise and their knowledge of SAS/OR® with other people.

All data files and sas codes used in this book are also available to download at www.sas-or.com. Solution to exercises are also avilable on this website.

Accessing the SAS/OR® Sample Library

The SAS/OR® sample library includes many examples that illustrate the use of SAS/OR® software. To access these sample programs, from the Help menu select "Learning to Use SAS," select "Sample SAS Programs," and choose "SAS/OR®" from the list of available topics.

Additional Documentation for SAS/OR®
Software as given in the sas.com

The main source for SAS/OR® is the *SAS/OR® User's Guide: Mathematical Programming,* which can be downloaded from www.sas.com.

Apart from this title, many other documents can be of help when using SAS/OR® software:

- *SAS/OR® User's Guide: Bills of Material Processing*

  This book provides documentation for the bill-of material (BOM) procedure and all BOM postprocessing SAS macros. The BOM procedure and SAS macros provide the ability to generate different reports and to perform several transactions that maintain and update BOMs

- *SAS/OR® User's Guide: Constraint Programming*

  This book provides documentation for the constraint programming procedure in SAS/OR® software and also serves as the primary documentation for the CLP (Constraint Linear Programming) procedure, an experimental procedure new to SAS/OR® software.

- *SAS/OR® User's Guide: Local Search Optimization*

  This book provides documentation for the local search optimization procedure in SAS/OR® software and also serves as the primary documentation for the genetic algorithm procedure, an experimental procedure that uses genetic algorithms to solve optimization problems.

- *SAS/OR® User's Guide: Project Management*

  This book provides documentation for the project management procedures in SAS/OR® software and also serves as the primary documentation for the CPM (A procedure for critical path method), DTREE (a procedure for decision tree analysis), NETDRAW (a procedure to draw a network), GANTT (a procedure to draw a GANTT chart), and PM (a procedure for project management), as well as PROJMAN application, a graphical user interface for project management.

- *SAS/OR® User's Guide: The QSIM Application*

  This book provides documentation for the QSIM (Queueing Simulation) Application, which is used to build and analyze models of queuing systems using discrete event simulation. This book shows you how to build models using the simple point-and-click graphical user interface, how to run the models, and how to collect and analyze the sample data to give you insight into the behavior of the system.

- *SAS/IRP® User's Guide: Inventory Replenishment Planning*

  This book provides documentation for SAS/IRP (Inventory Replenishment Planning) software. This book serves as the primary documentation for the IRP procedure for determining replenishment policies, as well as the %irpsim SAS programming macro for simulating replenishment policies.

Appendix 4: Syntax of
PROC NETFLOW*

A network consists of a collection of nodes joined by a collection of arcs. The arcs connect the nodes and convey flow of one or more commodities that are supplied at supply nodes and demanded at demand nodes in the network. Each arc has a cost per unit of flow, a flow capacity, and a lower flow bound associated with it. An important concept in network modeling is conservation of flow. Conservation of flow means the following:

(Total flow in arcs directed toward a node) + (Supply at the node) − (Demand at the node) = Total flow in arcs directed away from the node

Using this description, PROC NETFLOW accomplishes the following:

- Finds the flow through each arc in the network that minimizes the total cost of flow

- Meets the demand at demand nodes using the supply at supply nodes so that the flow through each arc is on or between the arc's lower flow bound and its capacity

- Satisfies the conservation of flow.

The general syntax of PROC NETFLOW is as follows:

```
PROC NETFLOW options;
CAPACITY variable;
COEF variables;
COLUMN variable;
CONOPT;
COST variable;
DEMAND variable;
HEADNODE variable;
ID variables;
LO variable;
NAME variable;
NODE variable;
PIVOT;
PRINT options;
RESET options;
ROW variables;
RHS variables;
SAVE options;
```

---

* For full details, see the *SAS/OR® User's Guide*.

```
SHOW options;
SUPDEM variable;
SUPPLY variable;
TAILNODE variable;
TYPE variable;
VAR variables;
QUIT;
RUN;
```

The options in PROC NETFLOW give the user very powerful tools that are summarized as follows:

- The NETFLOW procedure always creates and initializes a SAS macro called _ORNETFL at termination. After each PROC NETFLOW run, you can examine this macro by specifying %put _ORNETFL and see whether PROC NETFLOW ran correctly or what error or difficulty it encountered.

Appendix 5: Syntax of PROC CPM*

The CPM (Critical Path Method) procedure can be used for planning, controlling, and monitoring a project.

A typical project consists of several activities that may have precedence and time constraints. Some of these activities may already be in progress; some of them may follow different work schedules. All the activities may compete for scarce resources. PROC CPM enables the user to schedule activities subject to all these constraints.

The general syntax of PROC CPM is as follows:

```
PROC CPM options;
ACTIVITY variable;
ACTUAL / actual options;
ALIGNDATE variable;
ALIGNTYPE variable;
BASEL5INE baseline options;
CALID variable;
DURATION / duration options;
HEADNODE variable;
HOLIDAY variable / holiday options;
ID variables;
PROJECT variable / project options;
RESOURCE variables / resource options;
SUCCESSOR variables / lag options;
TAILNODE variable;
run;
```

- The CPM procedure defines a macro variable _ORCPM_. This variable contains a character string that indicates the status of the procedure.

---

* For full details, see the *SAS/OR® User's Guide.*

Appendix 6: Standard Normal Distribution

z	.00	.01	.02	.03	.04	.05	.06	.07	.08	.09
−3.9	.00005	.00005	.00004	.00004	.00004	.00004	.00004	.00004	.00003	.00003
−3.8	.00007	.00007	.00007	.00006	.00006	.00006	.00006	.00005	.00005	.00005
−3.7	.00011	.00010	.00010	.00010	.00009	.00009	.00008	.00008	.00008	.00008
−3.6	.00016	.00015	.00015	.00014	.00014	.00013	.00013	.00012	.00012	.00011
−3.5	.00023	.00022	.00022	.00021	.00020	.00019	.00019	.00018	.00017	.00017
−3.4	.00034	.00032	.00031	.00030	.00029	.00028	.00027	.00026	.00025	.00024
−3.3	.00048	.00047	.00045	.00043	.00042	.00040	.00039	.00038	.00036	.00035
−3.2	.00069	.00066	.00064	.00062	.00060	.00058	.00056	.00054	.00052	.00050
−3.1	.00097	.00094	.00090	.00087	.00084	.00082	.00079	.00076	.00074	.00071
−3.0	.00135	.00131	.00126	.00122	.00118	.00114	.00111	.00107	.00104	.00100
−2.9	.00187	.00181	.00175	.00169	.00164	.00159	.00154	.00149	.00144	.00139
−2.8	.00256	.00248	.00240	.00233	.00226	.00219	.00212	.00205	.00199	.00193
−2.7	.00347	.00336	.00326	.00317	.00307	.00298	.00289	.00280	.00272	.00264
−2.6	.00466	.00453	.00440	.00427	.00415	.00402	.00391	.00379	.00368	.00357
−2.5	.00621	.00604	.00587	.00570	.00554	.00539	.00523	.00508	.00494	.00480
−2.4	.00820	.00798	.00776	.00755	.00734	.00714	.00695	.00676	.00657	.00639
−2.3	.01072	.01044	.01017	.00990	.00964	.00939	.00914	.00889	.00866	.00842
−2.2	.01390	.01355	.01321	.01287	.01255	.01222	.01191	.01160	.01130	.01101
−2.1	.01786	.01743	.01700	.01659	.01618	.01578	.01539	.01500	.01463	.01426
−2.0	.02275	.02222	.02169	.02118	.02068	.02018	.01970	.01923	.01876	.01831
−1.9	.02872	.02807	.02743	.02680	.02619	.02559	.02500	.02442	.02385	.02330
−1.8	.03593	.03515	.03438	.03362	.03288	.03216	.03144	.03074	.03005	.02938
−1.7	.04457	.04363	.04272	.04182	.04093	.04006	.03920	.03836	.03754	.03673
−1.6	.05480	.05370	.05262	.05155	.05050	.04947	.04846	.04746	.04648	.04551
−1.5	.06681	.06552	.06426	.06301	.06178	.06057	.05938	.05821	.05705	.05592
−1.4	.08076	.07927	.07780	.07636	.07493	.07353	.07215	.07078	.06944	.06811
−1.3	.09680	.09510	.09342	.09176	.09012	.08851	.08691	.08534	.08379	.08226
−1.2	.11507	.11314	.11123	.10935	.10749	.10565	.10383	.10204	.10027	.09853
−1.1	.13567	.13350	.13136	.12924	.12714	.12507	.12302	.12100	.11900	.11702
−1.0	.15866	.15625	.15386	.15151	.14917	.14686	.14457	.14231	.14007	.13786
−0.9	.18406	.18141	.17879	.17619	.17361	.17106	.16853	.16602	.16354	.16109
−0.8	.21186	.20897	.20611	.20327	.20045	.19766	.19489	.19215	.18943	.18673
−0.7	.24196	.23885	.23576	.23270	.22965	.22663	.22363	.22065	.21770	.21476
−0.6	.27425	.27093	.26763	.26435	.26109	.25785	.25463	.25143	.24825	.24510
−0.5	.30854	.30503	.30153	.29806	.29460	.29116	.28774	.28434	.28096	.27760
−0.4	.34458	.34090	.33724	.33360	.32997	.32636	.32276	.31918	.31561	.31207
−0.3	.38209	.37828	.37448	.37070	.36693	.36317	.35942	.35569	.35197	.34827
−0.2	.42074	.41683	.41294	.40905	.40517	.40129	.39743	.39358	.38974	.38591
−0.1	.46017	.45620	.45224	.44828	.44433	.44038	.43644	.43251	.42858	.42465
−0.0	.50000	.49601	.49202	.48803	.48405	.48006	.47608	.47210	.46812	.46414

Appendix 6: Standard Normal Distribution

z	.00	.01	.02	.03	.04	.05	.06	.07	.08	.09
0.0	.50000	.50399	.50798	.51197	.51595	.51994	.52392	.52790	.53188	.53586
0.1	.53983	.54380	.54776	.55172	.55567	.55962	.56356	.56749	.57142	.57535
0.2	.57926	.58317	.58706	.59095	.59483	.59871	.60257	.60642	.61026	.61409
0.3	.61791	.62172	.62552	.62930	.63307	.63683	.64058	.64431	.64803	.65173
0.4	.65542	.65910	.66276	.66640	.67003	.67364	.67724	.68082	.68439	.68793
0.5	.69146	.69497	.69847	.70194	.70540	.70884	.71226	.71566	.71904	.72240
0.6	.72575	.72907	.73237	.73565	.73891	.74215	.74537	.74857	.75175	.75490
0.7	.75804	.76115	.76424	.76730	.77035	.77337	.77637	.77935	.78230	.78524
0.8	.78814	.79103	.79389	.79673	.79955	.80234	.80511	.80785	.81057	.81327
0.9	.81594	.81859	.82121	.82381	.82639	.82894	.83147	.83398	.83646	.83891
1.0	.84134	.84375	.84614	.84849	.85083	.85314	.85543	.85769	.85993	.86214
1.1	.86433	.86650	.86864	.87076	.87286	.87493	.87698	.87900	.88100	.88298
1.2	.88493	.88686	.88877	.89065	.89251	.89435	.89617	.89796	.89973	.90147
1.3	.90320	.90490	.90658	.90824	.90988	.91149	.91309	.91466	.91621	.91774
1.4	.91924	.92073	.92220	.92364	.92507	.92647	.92785	.92922	.93056	.93189
1.5	.93319	.93448	.93574	.93699	.93822	.93943	.94062	.94179	.94295	.94408
1.6	.94520	.94630	.94738	.94845	.94950	.95053	.95154	.95254	.95352	.95449
1.7	.95543	.95637	.95728	.95818	.95907	.95994	.96080	.96164	.96246	.96327
1.8	.96407	.96485	.96562	.96638	.96712	.96784	.96856	.96926	.96995	.97062
1.9	.97128	.97193	.97257	.97320	.97381	.97441	.97500	.97558	.97615	.97670
2.0	.97725	.97778	.97831	.97882	.97932	.97982	.98030	.98077	.98124	.98169
2.1	.98214	.98257	.98300	.98341	.98382	.98422	.98461	.98500	.98537	.98574
2.2	.98610	.98645	.98679	.98713	.98745	.98778	.98809	.98840	.98870	.98899
2.3	.98928	.98956	.98983	.99010	.99036	.99061	.99086	.99111	.99134	.99158
2.4	.99180	.99202	.99224	.99245	.99266	.99286	.99305	.99324	.99343	.99361
2.5	.99379	.99396	.99413	.99430	.99446	.99461	.99477	.99492	.99506	.99520
2.6	.99534	.99547	.99560	.99573	.99585	.99598	.99609	.99621	.99632	.99643
2.7	.99653	.99664	.99674	.99683	.99693	.99702	.99711	.99720	.99728	.99736
2.8	.99744	.99752	.99760	.99767	.99774	.99781	.99788	.99795	.99801	.99807
2.9	.99813	.99819	.99825	.99831	.99836	.99841	.99846	.99851	.99856	.99861
3.0	.99865	.99869	.99874	.99878	.99882	.99886	.99889	.99893	.99896	.99900
3.1	.99903	.99906	.99910	.99913	.99916	.99918	.99921	.99924	.99926	.99929
3.2	.99931	.99934	.99936	.99938	.99940	.99942	.99944	.99946	.99948	.99950
3.3	.99952	.99953	.99955	.99957	.99958	.99960	.99961	.99962	.99964	.99965
3.4	.99966	.99968	.99969	.99970	.99971	.99972	.99973	.99974	.99975	.99976
3.5	.99977	.99978	.99978	.99979	.99980	.99981	.99981	.99982	.99983	.99983
3.6	.99984	.99985	.99985	.99986	.99986	.99987	.99987	.99988	.99988	.99989
3.7	.99989	.99990	.99990	.99990	.99991	.99991	.99992	.99992	.99992	.99992
3.8	.99993	.99993	.99993	.99994	.99994	.99994	.99994	.99995	.99995	.99995
3.9	.99995	.99995	.99996	.99996	.99996	.99996	.99996	.99996	.99997	.99997

Table values represent area to the left of the z score.

References

Ağpak, K., and Gökçen, H. (2005). Assembly line balancing: two resource constrained cases. International Journal of Production Economics 96:129–140.

Amin, G. R., and Emrouznejad, A. (2006). An extended minimax disparity to determine the OWA operator weights. Computers and Industrial Engineering 50:312–316.

Amin, G. R., and Emrouznejad, A. (2009). Determining more realistic OWA weights. IEEE Fuzzy Systems and Knowledge Discovery 181:185.

Balas, E., and Ivanescu, P. L. (1964). On the generalized transportation problem. Management Science 11:188–202.

Castillo, E., Conejo, A. J., Pedregal, P., Garciá, R., and Alguacil, N. (2002). Building and Solving Mathematical Programming Models in Engineering and Science. Wiley, New York.

Charnes, A., and Cooper, W. W. (1961). Management Models and Industrial Applications of Linear Programming. Wiley, New York.

Charnes, A., Cooper, W., and Rhodes, E. (1978). "Measuring the efficiency of decision-making units," European Journal of Operational Research vol. 2, pp. 429–444.

Eisemann, K. (1964). The generalized stepping stone method for the machine loading model. Management Science 11:154–176.

Eklund, P., and Klawonn, F. (1992). Neural fuzzy logic programming. IEEE Transactions on Neural Networks 3(5):3815–3819.

Emrouznejad, A. (2005). Measurement efficiency and productivity in SAS/OR. Computers and Operations Research 32:1665–1683.

Emrouznejad, A. (2008). MP-OWA: The most preferred OWA operator. Knowledge-Based Systems 21:847–851.

Emrouznejad, A. (2010). SAS/OWA: Ordered weighted averaging in SAS optimization. Soft Computing 14:379–386.

Emrouznejad, A., and Amin., G. R. (2009). Document similarity: A new measure using OWA. IEEE Fuzzy Systems and Knowledge Discovery 186:190.

Emrouznejad, A., and Amin, G. R. (2010). Improving minimax disparity model to determine the OWA operator weights. Information Sciences 180:1477–1485.

Emrouznejad, A., and De Witte, K. (2010). COOPER-framework: A unified process for non-parametric projects. European Journal of Operational Research 207:1573–1586.

Emrouznejad, A., Parker, B. R., and Tavares, G. (2008). Evaluation of research in efficiency and productivity: A survey and analysis of the first 30 years of scholarly literature in DEA. Socio-Economic Planning Sciences 42:151–157.

Engemann K.J., Filev, D. P., and Yager, R. R. (1996). Modelling decision making using immediate probabilities. International Journal of General Systems, 24:281–294.

Färe, R., Grosskopf, S., Lindgren, B. & Roos, P. (1992). Productivity changes in Swedish pharmacies 1980–1989: a non-parametric Malmquist approach, Journal of Productivity Analysis, 3, pp. 85–101.

Färe, R., Grosskopf, S., Norris, M. & Zhang, Z. (1994). Productivity growth, technical progress, and efficiency change in industrialized countries, American Economic Review, 84, pp. 66–83.

Farrell, M.J. (1957) The measurement of productive efficiency, Journal of the Royal Statistical Society, Series A, General, 120, Part 3, pp. 253–281.

Ferguson, A. R., and Dantzig, G. B. (1956). The allocation of aircraft to routes—An example of linear programming under uncertain demand. Management Science 3:45–73.

Funkuyama, H. (1995), "Measuring efficiency and productivity growth in Japanese banking: a non-parametric frontier approach," Appl. Financial Econ., 5: 95–107.

Gökçen, H., and Erel, E. (1998). Binary integer formulation for mixed-model assembly line balancing problem. Computers and Industrial Engineering 34:451–461.

Ho, W. (2008). Integrated analytic hierarchy process and its applications – a literature review. European Journal of Operational Research 186:211–228.

Ho, W., Dey, P. K., and Higson, H. E. (2006). Multiple criteria decision making techniques in higher education. International Journal of Educational Management 20:319–337.

Ho, W., and Emrouznejad, A. (2009). Multi-criteria logistics distribution network design using SAS/OR. Expert Systems with Applications 36:7288–7298.

Ho, W., and Ji, P. (2003). Component scheduling for chip shooter machines: A hybrid genetic algorithm approach. Computers & Operations Research 30:2175–2189.

Ho, W., and Ji, P. (2004). A hybrid genetic algorithm for component sequencing and feeder arrangement. Journal of Intelligent Manufacturing 15:307–315.

Ho, W., and Ji, P. (2005), A genetic algorithm to optimise the component placement process in PCB assembly. International Journal of Advanced Manufacturing Technology 26:1397–1401.

Ho, W., and Ji, P. (2006a). A genetic algorithm approach to optimising component placement and retrieval sequence for chip shooter machines. International Journal of Advanced Manufacturing Technology 28:556–560.

Ho, W., and Ji, P. (2006b). Optimal Production Planning for PCB Assembly. Springer, London.

Jensen, P. A., and Bard, J. F. (2003). Operations Research: Models and Methods. Wiley, New York.

Ji, P., Wong, Y. S., Loh, H. T., and Lee, L. C. (1994). SMT production scheduling: A generalized transportation approach. International Journal of Production Research 32:2323–2333.

Kacprzy, J. 1990. Inductive learning from considerably erroneous examples with a specificity based stopping rule, Proceedings of International Conference on Fuzzy Logic and Neural Networks, Izuka, Japan, p. 819.

Lourie, J. R. (1964). Topology and computation of the generalized transportation problem. Management Science 11:177–187.

Patterson, J. H., and Albracht, J. J. (1975). Assembly line balancing: Zero-one programming with Fibonacci search. Operations Research 23:166–172.

Saaty, T. L. (1980). The Analytic Hierarchy Process. McGraw-Hill, New York.

SAS Institute Inc. (1990). SAS® Guide to Macro Processing, Version 6, Second Edition. Cary, NC: SAS Institute Inc.

SAS Institute Inc. (2004). SAS/IRP® 9.1.2 User's Guide: Inventory Replenishment Planning. Cary, NC: SAS Institute Inc.

SAS Institute Inc. (2004). SAS/OR ® 9.1 User's Guide: Project Management. Cary, NC: SAS Institute Inc.

SAS Institute Inc. (2010). SAS/OR ® 9.2 User's Guide: The QSIM Application. Cary, NC: SAS Institute Inc.

SAS Institute Inc. (2010). SAS/OR ® 9.22 User's Guide: Constraint Programming. Cary, NC: SAS Institute Inc.

SAS Institute Inc. (2010). SAS/OR ® 9.22 User's Guide: Local Search Optimization. Cary, NC: SAS Institute Inc.

SAS Institute Inc. (2010). SAS/OR ® 9.22 User's Guide: Bill of Material Processing. Cary, NC: SAS Institute Inc.

Taha, H. A. (2003). Operations Research: An Introduction. Prentice Hall, New Jersey.

Williams, H. P. (1999). Model-Building in Mathematical Programming. Wiley, New York.

Winston, W. L., and Venkataramanan, M. (2003). Introduction to Mathematical Programming: Operations Research. Brooks/Cole-Thomson Learning, California.

Yager, R. R. (1988). On ordered weighted averaging aggregation operators in multi-criteria decision making. IEEE Transactions on Systems, Man, and Cybernetics 18:183–190.

Yager, R. R. (1993). Families of OWA operators. Fuzzy Sets and Systems 59:125–148.

Yager, R. R., and Filev, D. P. (1992). Fuzzy logic controllers with flexible structures. In: Proc. Second Internat. Conf. on Fuzzy Sets and Neural Networks, pp. 317–320.

Yager, R. R., Kacprzyk, J., and Beliakov, G. (2011). Recent Developments in the Ordered Weighted Averaging Operators: Theory and Practice. Kluwer Academic Publishers, Norwell, Mass.

Index

Printed and bound by CPI Group (UK) Ltd, Croydon, CR0 4YY

24/10/2024

01778278-0010